Lighting the Electronic Office

Gary R. Steffy, IES, FIALD

VAN NOSTRAND REINHOLD

I(T)P™ A Division of International Thomson Publishing Inc.

New York ◆ Albany ◆ Bonn ◆ Boston ◆ Detroit ◆ London ◆ Madrid ◆ Melbourne
Mexico City ◆ Paris ◆ San Francisco ◆ Singapore ◆ Tokyo ◆ Toronto

Printed in the United States of America
For more information, contact:

Van Nostrand Reinhold
115 Fifth Avenue
New York, NY 10003

International Thomson Publishing GmbH
Königswinterer Strasse 418
53227 Bonn
Germany

International Thomson Publishing Europe
Berkshire House 168–173
High Holborn
London WCIV 7AA
England

International Thomson Publishing Asia
221 Henderson Road #05–10
Henderson Building
Singapore 0315

Thomas Nelson Australia
102 Dodds Street
South Melbourne, 3205
Victoria, Australia

International Thomson Publishing Japan
Hirakawacho Kyowa Building, 3F
2-2-1 Hirakawacho
Chiyoda-ku, 102 Tokyo
Japan

Nelson Canada
1120 Birchmount Road
Scarborough, Ontario
Canada M1K 5G4

International Thomson Editores
Campos Eliseos 385, Piso 7
Col. Polanco
11560 Mexico D.F. Mexico

1 2 3 4 5 6 7 8 9 10 COU-WF 01 00 99 98 97 96 95

Library of Congress Cataloging-in-Publication Data

Steffy, Gary R.
    Lighting the electronic office / Gary R. Steffy.
        p. cm.
    Includes bibliographical references and index.
    ISBN 0–442–01238–1
    1. Office buildings—Lighting. 2. Computer terminals—Lighting.
I. Title.
TK4399.035S74 1994
621.32'2523—dc20   94–23793
                                                   CIP

Contents

Preface

This book is intended to be a reference guide for architects, engineers, interior designers, facility managers, union administrators, and the workers interested in knowing more about lighting their work setting. Although there are quite a few voluntary guidelines and legislated standards regarding lighting in electronic workplaces around the world, no single resource surveys these guidelines and standards, reviews lighting criteria, proposes a model lighting guideline, and recommends methods of meeting criteria. Likewise, no single resource carefully and completely addresses the critical subject of lighting for electronic workplaces.

> ... lighting remains the single most important building system in determining how productive, functional, and comfortable people can be in electronic workplaces.

Too many times lighting is a chapter or a paragraph in a document that is intended to address all environmental and ergonomic concerns in the electronic workplace. As such, lighting does not receive full, complete, and correct coverage, yet remains the single most important building system in determining how comfortable and functional and ultimately how productive people can be in electronic workplaces.

Finally, as earth-friendly and user-friendly issues follow us into the twenty-first century, designers and managers will have an increasingly difficult task of balancing lighting quality, lighting quantity, and lighting energy to achieve comfort and productivity. This book will help designers and managers reach an appropriate equilibrium, but only if time is taken to understand the issues and address them in the design process. Though design processes can vary greatly from designer to designer and engineer to engineer, if the process includes addressing the issues identified in this text, then the lighting will better suit the people using computers.

As more and more legislative standards are developed, the people who design, manage, and administer electronic workplaces need to know what these standards mean and how to meet the criteria set forth within them. This book includes all of the following: an introduction to the issues; a discussion of the appropriate terms; current guidelines and standards, along with a review of their strengths and weaknesses; vision-related aspects, including lighting, that affect seeing in electronic workplaces, a model lighting guideline that should be considered in electronic workplaces; a discussion of available lighting equipment and techniques that can potentially meet criteria; several case studies; and a resource list.

With this reference, the reader will be in a position to create and/or evaluate better visibility work environments, resulting in improved user comfort and productivity.

Writing this book required many resources. All computer screen images illustrated in photographs are copyrighted by respective software developers/suppliers. All line art was developed by Gloria Paul. Much of the photography was done by Robert Eovaldi and Fred Golden. General Electric, Lithonia Lighting, Peerless Lighting, Gary Steffy Lighting Design Inc., and Van Nostrand Reinhold all provided valued contributions to this effort. Laura (my wife) and Heather (my daughter) have supported me in this effort for the past three years—thanks for being there always and for understanding. Thanks to Wendy Lochner, my editor, for encouraging the development of this book. Thanks to Laurie McGee the copy editor who provided invaluable guidance and asked some tough questions on terminology and phraseology.

My greatest debt, however, goes to the many clients over the past fifteen years who have had an open mind on lighting criteria issues and lighting resolutions so that their constituents (people using computers) can enjoy a better and more productive workplace.

Introduction

Many of the problems associated with working at video display terminal (VDT) screens are claimed to be due to physical limitations of the environmental setting and of the computer hardware. For example, neck and/or back pain associated with VDT work is generally blamed on inappropriate back support—the chair is at fault. Headaches are attributed to the characteristics of the VDT screen itself—background screen color, image clarity, flicker rate, and so on. Though there is no question that the screen may exacerbate headaches and that the chair may exacerbate neck and/or backaches, the fact remains that the VDT task is a *visual* task as well as a *manual* task. Headaches, backaches, and neckaches are the manifestations of the task of viewing or attempting to view the VDT screen and/or the associated paper documents. Bright reflections obscuring portions of the screen text or graphics may cause the observer to lean awkwardly in the chair. Although a better chair may make the awkward lean more comfortable, it does not and cannot eliminate the reason for the awkward lean—veiling images. This book introduces and clarifies the extent of the various vision and lighting issues; introduces the reader to various local, state, national and international lighting standards and guidelines for lighting electronic workplaces; proposes appropriate consensus criteria for improving the visibility of VDT screens and paper documents; proposes methods and/or equipment to meet this criteria; reviews various case studies illustrating appropriate lighting for electronic workplaces; and offers a resource list for obtaining various cited guidelines and lighting equipment.

> ... the VDT task is a *visual* task as well as a *manual* task.

Vision and lighting issues arising in the electronic workplace are addressed so that the reader understands the significance of these issues and can become more familiar with the terminology. This text establishes how interrelated are the tasks (the characteristics of the VDT screen and paper documents), the lighting (characteristics of surface reflectances, light levels and brightness levels, and contrasts), and the eyes (eye condition, age, and eye wear). These aspects are identified and reviewed so that the reader understands the scope of lighting for electronic work settings. Issues such as transient adaptation, accommodation, visual fatigue, distractions, veiling reflections, glare, and perceived light levels, among others, are introduced as key concepts.

A host of local, state, national, and international guidelines have been developed over the past decade to respond to the lack of appropriate lighting standards for electronic work settings and to force recognition of the interrelations previously described. These guidelines, while having a common goal of improving the electronic work setting for the computer user, in some instances have very different lighting-related criteria—depending on the kinds of visual work undertaken by the computer user. These differences are identified and discussed.

A model guideline for lighting for people using computers is proposed by the author that addresses both objective and subjective issues. Criteria in this proposed model guideline include task lighting levels (illuminances—both horizontal and vertical), ambient (general) lighting levels (both horizontal and vertical as well as uniformity), luminances (both absolute maximums and maximum-to-minimum ratios), surface reflectances, VDT-screen characteristics, subjective aspects, and others. The text discussion covers the rationale for these guidelines, and tabular formats are used for easy reference of information.

By example, lighting methods or techniques are introduced so that the reader is prepared to meet the model guidelines. Specific lighting equipment is identified that, when used properly, can meet the guideline criteria. To illustrate the effectiveness of the previously described consensus guideline and lighting techniques and equipment, a series of case studies is reviewed in detail. Project descriptions, lighting criteria summaries, lighting design ideas, and equipment selection and specification are all discussed. Black-and-white photos illustrate the resultant lighted electronic work settings.

Finally, a resource chapter includes a broad list of contact addresses for obtaining the key cited local, state, national, and international guidelines, and a contact list for the cited lighting equipment.

Biographical Data

Gary R. Steffy, IES, FIALD, founded his firm in 1982, after experiences as a lighting designer for the Detroit architectural engineering firm of Smith, Hinchman & Grylls and as a research engineer for Owens-Corning Fiberglass. His firm's work is global. His firm, Gary Steffy Lighting Design, Inc. in Ann Arbor, has provided lighting design consultation for The Prudential, GE, University of Michigan, Steelcase Inc., Sears, Land's End, Ford Motor Company, Alabama Power Company, and Pennsylvania Power & Light, among many others. Mr. Steffy spends more than half of his work day on the computer, typically using lighting calculation, word processing, financial, and graphics software as part of his work process.

Mr. Steffy is past president and Fellow of the International Association of Lighting Designers and is a member of the Illuminating Engineering Society of North America and of the International Commission on Illumination. He has published numerous articles, and his firm's work has been published in such publications as *Architectural Record, Interiors, Architecture, Architectural Lighting, Lighting Design & Application*, and *Interior Design*. His work on books includes coauthoring *Architectural Interior Systems* (VNR) and authoring *Architectural Lighting Design* (VNR). He was a founding director of the National Council on the Qualifications of the Lighting Professions. In 1994, he passed the inaugural Illuminating Engineering Society's Technical Knowledge Exam and received the IES Certificate of Technical Knowledge.

Mr. Steffy, a native of Lancaster, Pennsylvania, graduated from The Pennsylvania State University with a bachelor's degree in Architectural Engineering. He lives with his wife, Laura, and daughter, Heather, in Ann Arbor, Michigan.

Conventions

An attempt has been made to report units of measure consistently in SI convention throughout this text. Globally, with the exception of the European Community, there is a tendency to mix SI and U.S. Customary units. Further complications arise when dealing with soft and hard metric conversions and when "product standard dimensions" are not reported in standardized fashion globally. For example, 48-inch fluorescent lamps are 48 inches from the end of the lamp pins at one end to the end of the lamp pins at the other end. From the socket cap at one end to the socket cap at the other end, however, the lamps measure nearly, but not exactly, 1200 millimeters (mm). Hence, in one market the lamp is a "metric" lamp of 1200 mm length, and in another market the identical lamp is a "U.S. customary" lamp of 48-inch length. The reader is advised to beware how various manufacturers' data and various guidelines' recommendations are reported and/or whether soft (rounded) or hard (exact) measurement techniques are used. For convenience, the following table of measures is offered:

Measure	U.S. Customary	Systeme Internationale (SI)
Length	1 inch (in)	25.4 millimeters (mm)
	1 foot (ft)	305 mm
Illuminance	1 footcandle (fc)	10.76 lux (lx)
Luminance	1 footLambert (fL)	3.4 candelas/square meter (cd/m^2)

As much as practical, hard SI values are used when lighting criteria is presented in this text. This helps minimize inaccuracies arising from the rounding of soft conversions, particularly over several iterations. Therefore, when comparing specific criteria data with other references using soft conversions there may be discrepancies. For example, the Illuminating Engineering Society of North America (IESNA) uses the approximation of 10 lux to 1 footcandle, a discrepancy of nearly 10 percent. If the IESNA data is converted several times from U.S. Customary to SI using soft conversions in some cases and hard conversions in others, then the discrepancy in criteria values can easily grow to 20 or 30 percent.

Lighting data referenced in *Using the Model Guideline: Case Studies* were calculated using industry-recognized software, Lumen-Micro 6.01® by Lighting Technologies™ of Boulder, Colorado, USA, and/or were measured using Minolta™ 1° luminance meter (for luminances) and Minolta T-1™ illuminance meter.

Lighting the Electronic Office

1 | Computers, Lighting, and People

Computers have been around since the mid-1940s.[1] Until the mid- to late 1970s, however, only mainframe computers (large, centrally located machines) almost exclusively performed tedious number-crunching tasks. Large corporations involved in extensive research and development based on statistics and numerical analyses used computers. Only a handful of people—some engineers, scientists, and mathematicians—used computers. Even then, interaction between computer and human was primarily achieved with punch cards as input, essentially a classic typewriting-type of visual task, and paper printouts as output, essentially a poor-contrast reading task because ribbons tended to dry out easily.

By the mid-1970s, service businesses like banking, travel agencies, and insurance companies saw significant benefits to using computers to do more than *just* research or accounting—computers could track "events." Further, computers could store data for future reference and manipulation. Computers could improve the pace, appearance, and accuracy of written communications. Documents could be composed, revised, reformatted, and corrected without retyping the entire document.

Nevertheless, computers were still expensive and finicky devices that were used by few people until the early 1980s. In January 1991 *The Harvard Business Review* reported that just twenty years before (1971), fewer than fifty thousand computers altogether were in use, yet by 1991, fifty thousand computers were purchased *each day.* The IBM Personal Computer® changed the office landscape forever in 1981.[1] Any task, whether requiring actual computing power ("number crunching"), mass data storage, or unprecedented presentation quality, could be, was, and is today performed on a computer by most every person in the work envi-

Why do we continue to overlight spaces?

1

ronment. Office clerks; secretaries; executives; designers; engineers; accountants; airline, car, train, and hotel reservationists; insurance agents; telephone operators; drafters; cashiers (retail, fast-food, restaurant, grocery); tellers; brokers; mail-order operators; students; librarians; and so on find the computer as integral to their work as pencils and paper once had been.

This proliferation of computers has wrought a host of problems, generally categorized as "ergonomic" issues—that is, issues related to the physiological and psychological needs of humans. Neckaches, headaches, wrist ailments (e.g., carpal tunnel syndrome), and visual fatigue are several such issues. This book addresses vision-related issues, specifically dwelling on lighting aspects related to computer use by office workers. Recognize, however, that these ergonomic problems are quite interrelated.

> Lighting can be significantly more effective, more comfortable, and more efficient if *luminance* issues are addressed in addition to *illuminance* (light level) issues.

Indeed it is reasonable to assume that such problems as neckaches and headaches are not due entirely, if at all, to those elements commonly blamed—immobile work surfaces, nonadjusting or poorly designed chairs, video display monitors, and so on. Some of these problems likely result from, or at least are compounded by, inappropriate lighting (daylight and/or electric light).

To a large extent, inappropriate lighting is the result of historical attitudes and design methods. Throughout the 1950s, 1960s, and the early 1970s, energy was cheap and "limitless." Building architecture evolved as monuments to mass production, modular construction techniques. Many buildings in the northern United States required more heating than cooling, and lighting was seen as a means of augmenting the heating. Handwritten paper tasks were the norm. Mimeograph and early xerograph copies had inherently low contrast, thereby requiring high light levels for people to read them fast and accurately. "More light, better sight" was the slogan of the era. Of course, many of the personal computers purchased by the mid-1980s were installed in buildings built for another era. Even today, however, many lighting decisions are based on the old slogan, more light, better sight. Indeed, by the late 1970s it was clear that most people (with rather "average" eyesight) could comfortably and accurately read handwritten, typewritten, published, xerographed material with less than 320 lux of light. Why then do we continue to overlight workspaces to 800 lux?

Although most people can read typical printed material quite well with just 100 to 215 lux[5] on the paperwork, a room with 100 to 215 lux on the task area might appear dim if no other lighting criteria are considered. Herein lies the problem with lighting design. Designing to lux or illuminance levels on anticipated task areas can only ensure that enough light exists under which most people could potentially perform tasks. Designing to lux or illuminance levels does nothing to ensure that there will be appropriate task brightness-to-room surface brightness ratios to make the task easily visible or that there will be adequate surface brightnesses to make the work area comfortable and pleasing—"comfortable" meaning appropriate brightness balancing between the task area(s) and the surrounding areas to allow for comfortable adaptation and to minimize glare. "Pleasing" meaning appropriate brightness patterning throughout the space to meet subjective or psychological needs. Lighting can be significantly more effective, more comfortable, and more efficient if brightness issues (technically known as luminances) are addressed in addition to illuminance issues. As Grossman's Law represents, "complex problems have simple, easy-to-understand wrong answers." Indeed, the only simple way— read "less design time and fewer lights to buy and install"—to get a room to appear brighter is to simply use more lamps in each luminaire (light fixture) and therefore throw more lux onto the work area; maybe by accident enough light will bounce around onto the walls and ceilings and thereby create a brighter-looking space. This wastes energy, creates havoc with the viewing of computer monitors, and in all likelihood even makes the paper tasks less visible because of increased glare and veiling reflections. Nevertheless, this is how we have been designing lighting for more than thirty years! With the advent of computers and with the rising concern over the waste of environmental resources, this design approach should no longer be tolerated. Yet rather than accept the lighting system design as the culprit, we put blame on chairs, desks, computers, and workers.

Why haven't reading and lighting difficulties come up in the past? Probably because no task is nearly so demanding on so many people as is reading a computer screen. Reading paper documents, though seemingly quite similar to reading a computer screen, is a lot more flexible and forgiving than reading the VDT (video display terminal) screen. The very nature of the paper task allows us to hold the paper at various angles to eliminate or minimize harsh light reflections and/or change viewing distance quite readily to ease accommodation (focusing). Additionally, the ways of operating in the old "paper-full" office are different than the ways of operating in a "paper-less" office. Consider the following two scenarios:

... buying chairs and furniture accessories without changing poor lighting is treating the symptom and not the illness.

Scenario 1: It is 1973 in an accounts-receivable office somewhere. People are recording payments received on outstanding invoices. A box of mail is delivered to a desk. The individual at the desk performs one set of specific tasks and then turns the "result" over to another person, who will perform a set of tasks, who then turns the result over to another individual who performs another set of specific tasks. Essentially an assembly line approach is taken, with each person performing one aspect of the task. The people in this group constitute accounts receivable and collectively perform the work of receiving payments. The tasks are something like this:

▼ Individual #1 opens the mail and sorts it into payments received and inquiries—note the variety of visual assessments necessary. The individual reads writing on an envelope, slits the envelope with the letter opener, empties the envelope, likely takes an inked rubber stamp and stamps the material "Received," filling in the date of receipt, and stacks the material into at least two different stacks—payments and inquiries. The reading portion of this individual's task is relatively minimal and is interrupted with significant amounts of other work (slitting the envelope, stamping "Received," writing down a receipt date, placing onto a stack).

▼ Individual #2 then takes the payments received, and reads each check for proper payer information and to assess the amount tendered—this is a reading task (reading pen, typewritten, laser-printed, and/or bonding-print). The amount received and date of receipt are then entered into a log book—this is a handwritten task. Recognize, however, that this handwritten task is of short duration. After each log entry, the individual "breaks" from the handwriting (and reading) task to place the "remittance" copy into a basket full of remittance copies and then reaches for another payment.

▼ Individual #3 then takes the remittance copies of the invoices that were returned along with the payment and files these under the names (or numbers) of the respective accounts. Again, the visual portion of the work is interrupted by other tasks—the activity of actually filing; reaching for another copy to file; going back to individual #2's desk to pick up another basket of remittance copies.

▼ Individual #4 takes the inquiries that individual #1 sorted and pulls together material for responses. This likely includes going to the files and searching for previous remittance copies or perhaps going to another department to find the status of an order.

Scenario 2: It is 1993 in an accounts-receivable office somewhere. People are recording payments received on outstanding invoices. Now, however, the tasks are something like this:

▼ Each individual in the group performs all of the tasks outlined in scenario 1 and previously performed by at least four different individuals. This can be achieved only because of the technology of the personal computer (PC). Now, with the help of the computer, the individual need only: (1) open envelopes; (2) visually scan the remittance copy for the client's account number and amount due; (3) read the enclosed check for proper payer information and assess amount tendered; (4) enter amount received onto the electronic ledger; (5) if/as applicable, address inquiries by "calling up" the client's account name/number on the electronic ledger, "conversing" with the computer, and word processing a response. Recognize that the visual task time is now more continuous, with fewer interruptions of nonvisual aspects. Reading becomes a bigger portion of the work pie! Problems that previously went unnoticed by workers, or about which complaints were not made because the problems didn't seem significant enough with regard to the whole work assignment, are now much more likely to result in complaints and/or reduced comfort and ultimately reduced productivity.

In other words, the way in which we do work has changed significantly over the past ten years, thanks to the technology of the PC in conjunction with techniques to make businesses more competitive.[2] We now find ourselves sitting at computers incessantly, reading video monitors for relatively long periods of time. This very sedentary, visually demanding work now magnifies all that was (and still is) wrong with many of our work environments.

Figure 1-1 is a close-up view of a VDT screen with significant patches of intense light images obliterating or veiling portions of the screen text. An observer can "miss" seeing some or all of these veiling images if he or she slumps in the chair in a sideways fashion or crooks his or her neck. This is a condition that can be lessened with an adjustable VDT stand *combined with* an adjustable chair. The effectiveness of a VDT stand alone, however, is questionable because of the limits of flexibility of the stand versus the viewing height and posture preferences of the seated observer. An "ergonomically correct" chair alone probably will not dramatically improve viewing conditions or entirely rid the observer of the physical ailments manifested by poor lighting conditions. In other words, buying ergonomically correct chairs for people and adjustable furniture accessories for the VDT without changing poor lighting and/or poor task conditions is treating the symptom and not the illness. The condition shown in Figure 1-1 can be dramatically minimized, if not eliminated, with appropriate lighting and/or with an optically improved VDT screen.

Today's computer users do not use the computer exclusively to perform work tasks. Paper-based reference materials are frequently used (Figure 1-2). Some folks spend a good bit of time reading hard-copy documents and refer to the computer

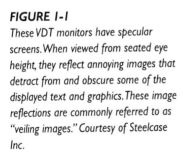

FIGURE 1-1
These VDT monitors have specular screens. When viewed from seated eye height, they reflect annoying images that detract from and obscure some of the displayed text and graphics. These image reflections are commonly referred to as "veiling images." Courtesy of Steelcase Inc.

FIGURE 1-2
Many VDTs are used in conjunction with other tasks. Reading paper documents may comprise a larger portion of time and/or may be a more significant component of productivity. Courtesy General Electric.

on an infrequent basis. Other people do use the computer as the sole source of visual information during the course of a workday—conversing with the computer to accomplish their work. This great diversity in visual tasks demands that great care be taken in developing lighting solutions. A solution that may work well for one sort of task may be totally inappropriate for another sort of task. Indeed, herein lies the problem. Many of the environments in which VDTs are found were designed with lighting approaches appropriate mostly or solely for viewing paper tasks. In most cases the quantity and quality of light required for paper tasks are entirely different than that required for VDT viewing.

Defining the Problem

There is no panacea—no lighting solution will solve all of the problems all of the time. On the other hand, however, one need not be a rocket scientist to develop better lighting conditions for people using VDTs. Common sense goes a long way. A careful review of potential problems and potential solutions can generally result in acceptable, comfortable electronic work environments. Essentially, three basic criteria groups need to be addressed when assessing electronic work environments. The three together appear, at times, to be mutually exclusive! These three basic criteria groups are ***economic ("purchaser-friendly")***, ***earth-friendly***, and ***user-friendly***.

The successful designer needs to "uncover" the appropriate criteria in each grouping, prioritize the criteria with the client, and develop solutions that address these criteria. In other words, the general problem of design means balancing the three criteria groups and educating the client and/or end user about the benefits and/or disadvantages involved in meeting or not meeting the various criteria. This text primarily addresses *user-friendly* issues and, secondarily, *earth-friendly* issues. Economic issues are not addressed specifically in terms of equipment costs and operating costs, because these are only a part of the economics that must be considered.

If only equipment and installation first costs and operating costs are considered, then user-friendly and earth-friendly issues are moot. Yet if first costs and operating costs are addressed to the exclusion of user-friendly and earth-friendly issues, it is likely that ultimate costs in terms of salaries, benefits, hiring, and training programs will be very high relative to productivity. In some cases, such costs can simply be passed along to society in the form of higher product and service costs and/or in higher landfill and environmental protection costs. In other cases, firms may literally go out of business because employee satisfaction and productivity are unacceptably and inexplicably low; in other words, the business is not competitive on a long-term basis.

> . . . legislation and social pressure will push electronic office environments toward more comfortable, productive conditions.

Defining the problem is the most difficult part of any commission. Often it is hastily defined and is based on archaic definitions that mask the real problem(s) and challenge(s) in design. To better understand and, more important, to better implement appropriate VDT environments, it is necessary to define the problem. The problem is maximizing task visibility and comfort, thereby presumably increasing productivity of people. Never lose sight of the problem. Designing space and lighting solutions around the computer hardware will not solve people problems. Indeed, people will take matters into their own hands—turning off lights, delamping luminaires (actually removing lamps), putting up cardboard or paper shades (over windows) or cardboard glare shields (around computer screens or offending lights).

It is considered by many to be cheapest and easiest to plop new technology into existing environments and let occupants adapt as best they can. The cheapest chairs are those already in place. The cheaper *new* chairs are the least forgiving ergonomically. The cheapest work surfaces are those already in place. The cheaper *new* work surfaces are the least forgiving ergonomically. The cheapest lighting is that which is already in place. The cheaper *new* lighting is 600 mm by 1200 mm, 3- and 4-lamp, lensed luminaires. Of course, all of this cheap stuff becomes cheaper still because more and more is purchased over time because it is the

> Workers who are more comfortable and satisfied are more likely to:
> ▼ Stay at the job longer,
> ▼ Require less down time for breaks,
> ▼ Have fewer errors.
> ▼ Take fewer absences . . . in other words,
> ▼ Be more productive!

cheapest. The money-making devices of capitalism encourage mass production from tooling that has long since been expensed.

Generally such mass-marketing successes die only when legislated out of existence or when social pressure reaches substantial levels. Both forms of market manipulation (legislation and social pressure) are pushing electronic work environments toward more comfortable conditions, which will likely have the side benefit of increasing productivity. Indeed, money must be spent to make money. Productivity improvements alone can pay back lighting equipment investments in rather short order. Many times the short-term view of "how much will it cost today" masks the long-term view of "how much can be made on the investment." Workers who are more comfortable and satisfied are more likely to: stay at the job longer, have less down time for breaks, require less down time for training, make fewer errors, and be absent less due to perceived and real ills associated with poor viewing conditions (headaches, visual fatigue, neckaches, and so on).

Unfortunately, the productivity gains associated with improved lighting conditions can only be based on anecdotal evidence rather than on research. A significant amount of research needs to be undertaken to quantify the productivity gains associated with "user-friendly" lighting. Nevertheless, the economic significance of improved productivity is too great to ignore entirely. It has been suggested that improved lighting may increase productivity by 1 percent.[3] Even if productivity increases by 0.25 percent, the impact on payback can be significant. In other words, proper new and retrofit lighting systems can be shown to be purchaser-friendly if the economic story encompasses more than just initial costs.

Recent advances in lamp, ballast, and luminaire technologies offer significant opportunities for energy efficiency if implemented properly. Electric utilities' desires to capture electrical capacity from the existing customer base rather than build new power plants has resulted in rebate and financing programs of the retrofit of lighting systems. Lighting systems that are more sympathetic to the earth's environment are now available and affordable in many regions where high energy rates and significant demand-factor penalties are commonplace. Couple energy savings with productivity improvements, and proper new or retrofit lighting systems can offer fast paybacks. Even where energy rates are low, life-cycle costs typically favor energy-prudent solutions.

Why isn't proper lighting developed and installed in more electronic work environments? More appropriate lighting criteria, which offer specific guidance toward developing user-friendly environments, have been developed, researched, and applied over the past decade. Many of these criteria are technically based, however, and many times either not known or not understood by the designers of the lighting, let alone by the purchasers of the lighting or the workers using the lighting. At other times, only the objective criteria are applied without consideration of the more subjective issues of preference and comfort. As purchasers, users, designers, legislators, and others become more familiar with the issues of design for the electronic workplace, more appropriate designs will evolve and be implemented more frequently. The remainder of this text is devoted to educating people on the lighting aspects of electronic workplaces.

Before explicitly reviewing lighting criteria for the electronic workplace, some key concepts are presented. Then a review is made of the basic elements that are responsible for the success of lighting. With an understanding of these key concepts and basic elements, the criteria guidelines later discussed will be more readily obvious as important and necessary to successful lighting.

Key Concepts

Solving for deficiencies in the electronic work environment is difficult to accomplish without understanding the problems. To know and identify these problems requires some knowledge of the ergonomic issues related to vision and subsequently to lighting. Some key concepts are outlined briefly in this section. A variety of references were used as source material for some of the following concept definitions, and more definitive, technically accurate explanations can be found in the Illuminating Engineering Society of North America's *Lighting Handbook Reference & Application,*[4] *Human Factors in Lighting,*[5] and *Architectural Lighting Design.*[6]

Accommodation. The process by which the eyes change focus. This is a particularly important concept, since looking from reference paper documents on a horizontal work surface to a vertical VDT screen may involve different viewing distances (or focal lengths) and thereby require accommodation. Figure 1-3 graphically depicts the varying viewing distances that are found in many typical

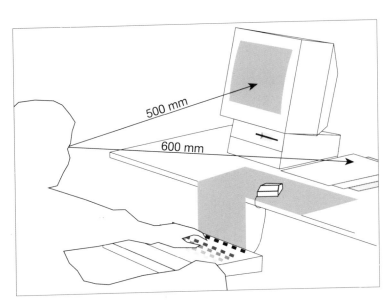

FIGURE 1-3

Accommodation for the viewer not using a paper-document holder varies as the viewer looks between the VDT screen image, which might be 500 mm away from the eyes, and the paper, which might be 600 mm away from the eyes.

VDT workspaces; Figure 1-4 depicts an appropriate arrangement of paper on a paper-document holder and nearby VDT. Constant viewing between the paper and VDT screen and the subsequent accommodation can cause visual fatigue. *Place the reference or source paper document at the same distance from and at the same elevation as the VDT screen with a paper-document holder.*

Adaptation. The process by which the eyes adjust from one luminance or color intensity to another. Prolonged viewing at one luminance or color intensity can "saturate" the viewing system, requiring a period of "recovery" (adaptation) to another luminance or color intensity. As with accommodation, adaptation is a particularly important concept with respect to task performance in the electronic office. Viewing a VDT screen with little text/graphic imagery on a dark screen background (positive-contrast screen) and then changing view to a lighted piece of white or light-colored paper can result in discomfort. On the other hand, viewing a lighted piece of white or light-colored paper and then changing view to a VDT screen with little text-graphic imagery on a dark screen background will likely require some short time frame of adaptation prior to successful/quick task performance. Generally, lighter background screens (negative contrast screens) and matte-finish screens can reduce the adaptation effects experienced when alternating view from paper documents to VDT screen.

Ambient Lighting. General, overall lighting throughout a space is sometimes called ambient lighting. Ambient lighting can be direct, indirect, or a combination.

Brightness. The subjective assessment of reflected and transmitted light. Brightness is most closely associated with the measurable, objective quantity of ***luminance***, which is the quantity of reflected or transmitted light from an object/surface in a

FIGURE 1-4

When the paper document is at the same elevation and distance from the eyes as is the VDT screen image, accommodation is lessened, thereby minimizing visual fatigue effects associated with accommodation.

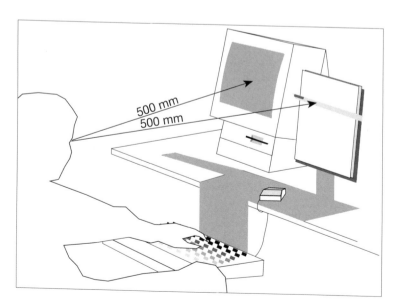

500 mm
500 mm

specific direction of view *as detected by a meter*. Brightness, however, depends on the conditions of the observers' eyes—brightness is the reflected or transmitted light *as detected by the eyes*. What is comfortable to some people may be "bright" to others and even "glary" to still others. Brightness, however, is a phenomenon that is relative to surrounding light conditions. For example, car headlights on a rainy night may be "too bright" or "glary" to most drivers. The surrounding conditions are of darkness (very low brightness), hence the observers' eyes are at a low or dark adaptation level. Any bright object or surface (e.g., car headlights at night) introduced into observers' fields of view will be seen as quite bright compared to their dark-adapted state. Yet during the day, car headlights may actually seem "dim" to many people. This relative condition is critical in the design of lighting in any space VDTs are used. Generally, the fewer "bright spots" that are created in the space, the less likely are complaints about glare and VDT screen veiling reflections. Minimizing bright spots means limiting, if not eliminating, surfaces of widely different reflectance or transmittance values; limiting or eliminating widely varying illuminances; limiting or eliminating direct lamp brightness; and/or creating softly glowing background conditions against which somewhat brighter zones or tasks appear less harsh.

CADD Task. Computer-aided design and drafting (CADD) tasks are common in many of the architectural, graphics, engineering, and industrial design fields. These CADD tasks generally include both computer screen viewing and paper document viewing (drawings, manuals, and so on). As such, the CADD task is likely the most difficult task for which to design lighting, particularly if the "old school" of designing to very high illuminances on drafting surfaces is used as a basis for design. The level of drawing detail and the inherent paper task contrast (e.g., crisp, high-contrast ink-drawn originals versus blurred, low-contrast blueline prints) should help guide the light levels required for reading. It is unlikely that hundreds of footcandles of light are necessary. The one exception to this would be the reading of drawings done on metal plates—inscribed lines on metal surfaces, which were used in the past for automotive drawings—which may require very high light levels and/or very specifically aimed lighting.

Chromatic Contrast. The color difference between two or more markings, areas, surfaces, or objects. Chromatic contrast and **luminance contrast** are responsible for what, how well, and how quickly we see. A light-blue fabric may reflect as much light as a light-yellow fabric, hence luminance contrast (brightness difference) between the two fabrics is zero. Yet we distinguish the difference between the two fabrics by the color difference or *chromatic contrast.*

Color Rendering. How well the light emitted from a manmade or natural light source renders colors of furnishings, clothing, artwork, skin, etcetera; generally reported on a scale where "100" represents optimal "true" color rendering. "True" is in quotes because the scale is based on a grouping of rather arbitrary colors established in the early 1960s. A reference or base condition of an artificially created "daylight" is the condition against which all other electric light sources are measured. The concept of color rendering is critical to chromatic contrast and the general "crispness" or "vividness" of an environment. Color contrast is diminished when using poor or lower-color-rendering lamps (typically lamps with color-rendering indices less than 65). Color contrast and color rendering is best judged by reviewing actual operating samples of lamps, rather than depending on the color rendering index (CRI). *The newer family of triphosphor fluorescent lamps have the best color-rendering properties for the money and are also the most efficient lamps available for interior applications.*

Color Temperature. The color of the light emitted by a lamp measured in degrees Kelvin (°K). Light of 2800°K to 3200°K is warm toned, and 3500°K is neutral white. Light of 3700°K to 4200°K is cool toned, and 5400°K is blue-white. There is little substantive research to support a specific color temperature recommendation for a given application. This tends to be a personal preference issue. Very low color temperature lamps (below 2500°K), however, promote a rather dim, yellow, candle-lit appearance; very high color temperature lamps (above 5000°K) promote a rather cool, gloomy, blue appearance. *Experience indicates that a warm-white-to-neutral-white color temperature* (2500°K to 3700°K) tends to be more complementary to skintones and more preferable to many users.

Conversational VDT Task. Primary task is "conversing" with the computer; hence, visual task is predominantly viewing the screen.

Cutoff Angle. The angle above which direct viewing of the lamp is not possible. At angles less than the cutoff angle, direct viewing of the lamp is possible and, therefore, glare potential increases. Technically this angle is defined as the angle between luminaire nadir (straight down) and the line of sight at which the lamp is not visible.

Data-Entry VDT Task. Primary task is "entering" data or information into the computer; hence, visual task is likely to be viewing paperwork.

Distractions. Caused by people walking behind the worker with the subsequent images of walking people silhouetting onto the VDT screen. *Distractions can be avoided with appropriate VDT orientation and/or use of 60-inch or higher workstation partitions to shield the VDT screen from surrounding activity.*

Dot-Pitch. A measure of VDT screen sharpness, reported in millimeters (mm). The smaller or lower the dot-pitch, the better the screen sharpness. Screen sharpness also depends on the kind of internal graphics hardware of the computer. Without the best monitor dot-pitch, however, internal graphics hardware cannot improve sharpness of image or text on a screen. A dot-pitch of 0.25 mm is considered best for most screen sizes, although the smaller the monitor, the smaller the dot-pitch must be in order to maintain screen sharpness equivalent to larger monitors. *For 14-, 15-, and 17-inch monitors, consider those with dot-pitch of 0.25 mm or smaller. The more detailed and/or lengthy the visual work at the monitor, consider smaller dot-pitch.*

Electromagnetic Ballasts. Essentially, an iron core surrounded by a coil of wire used to start and operate fluorescent lamps. Electromagnetic ballasts have been the industry standard since the late 1930s and are considered very reliable. Operating at 60 Hertz (Hz), these ballasts do have a tendency to have a slight hum and may cause noticeable flicker for some people. Although electromagnetic ballasts are available in premium energy-saving versions, the electronic ballast is the most efficient ballast available. Electromagnetic ballasts typically weigh 1.3 kilograms, generate additional heat above and beyond lamps' heat and require 10 percent to 20 percent additional energy above and beyond the energy required for the lamps. These typically should no longer be used and should be replaced by electronic ballasts for efficient, flicker-free, noise-free operation.

Electronic Ballasts. Ballasts that have been available since about 1982 but have generally been considered both expensive (without payback) and unreliable. Although some so-called electronic ballasts are actually hybrid electromagnetic ballasts, some of the newer and better electronic ballasts are essentially all-transistor components. Operating at 30,000 Hz, these ballasts are inaudible and flicker-free. The better electronic ballasts also provide lamp current-crest factors of 1.7 or less (within lamp manufacturers' recommendations), which assists in achieving rated lamp life; provide power factors approaching 1.0, which assists in achieving high electrical system/transmission efficiency; and low electromagnetic interference (EMI) with other electrical devices. *Consider only using electronic ballasts for fluorescent lighting solutions.*

Energy. The amount of power, measured in kilowatt-hours, used over a period of time. Energy is the definitive measure of electricity consumption rather than watts or watts per square meter, which only conveys connected load, not actual electricity use.

Full-Spectrum Lighting. Generally, "full-spectrum lighting" is a misnomer, more related to the marketing savvy of some fluorescent lamp salespeople than to any significant spectral difference from a variety of "standard" lamps commonly available. Specifically, reports that claim that full-spectrum lighting has been experienced by "many" people and "preferred" by those many people are commonplace. Unfortunately, substantive, repeatable research in this area has not been forthcoming. Indeed, the United States federal government funded some research work in the late 1980s at Lawrence Berkeley Laboratories that essentially found no grounds for promotion of full-spectrum lighting as better or more beneficial than any other lighting. Since full-spectrum lamps are generally about 30 percent less efficient than standard lamps, they are less bright. Hence, in existing VDT environments, where old, bright, glary fluorescent lights cause problems, retrofitting with the full-spectrum lamps can actually improve visual conditions. Recognize, however, that these lamps use as much wattage as standard lamps. In other words, the same benefits of less glare are likely to be achieved by delamping (taking out lamps) and using the more efficient "standard" fluorescent lamps which also cost about one-fifth the cost of full-spectrum fluorescent lamps. Finally, many full-spectrum lamps typically have a blue cast, resulting in a gloomy, overcast appearance. In short, full-spectrum lamps appear to be expensive, inefficient placebos.

Glare. Bright reflections from the VDT screen that cause squinting and obliteration of the VDT screen text/graphics (reflected glare); brightness directly from electric lights and/or daylighting that causes discomfort or disables or partially disables the worker from performing visual work. Glare is perhaps the most significant and obvious problem in the electronic office that is related to lighting. Some objective lighting criteria (e.g., luminances, luminance ratios) are directly related to minimizing, if not eliminating glare. Forfeiting some or all of this objective lighting criteria will likely result in glare.

High Light Levels. Light levels that are "too high" for the preference of some workers. This condition is generally related more to the excess brightness of room surfaces, the glare from the VDT, or veiling reflections on the VDT than to actual general light levels. Most people equate glare and/or excessive brightness to high light levels or too much light, hence misidentifying a very real problem. Sometimes this leads to addressing the symptoms and not the illness, like switching off lights or taking out some of the lamps in each luminaire rather than introducing appropriate lighting equipment. This can actually lead to continued visual fatigue, tired eyes, and/or headaches from attempting to read hard-copy text or line drawing detail in the lower light levels. The solution to "light levels are too high" is not necessarily simply reducing light levels. *Assess the viewing conditions and tasks and ascertain the actual problem before recommending solutions.*

Illuminance. The quantity of light falling on a specific task, surface or area, measured in lux (the English unit is footcandle, with 1 fc the equivalent of 10.76 lux). Illuminances are only partially responsible for how well people see. The interaction of illuminance with a surface results in either reflected and/or transmitted light. Hence, the surface that is illuminated has as much, if not a greater, influence on how well people see as does illuminance.

Interlaced Monitors. Only half of the screen is refreshed at a time on these VDT screens (every other line of the screen is refreshed electronically top to bottom first and then the refresh process begins again at the top by refreshing the remaining intermediate lines). This typically causes significant flicker. *See* **noninterlaced monitors.** *Only consider noninterlaced monitors.*

Low Light Levels. Levels of light that are "too low" for the preference of some workers. This condition is generally related more to the low brightness of room surfaces and/or the poor visibility of paper documents than to actual general light levels. Many times there is sufficient light on the paper task to provide good task contrast. If, however, walls and ceilings are quite dark relative to the paper task, then two conditions are likely to bother the observer, if not cause visual fatigue: adaptation effects (viewing from the bright paper to the darker surroundings) and general impressions of "dinginess" or "darkness," neither of which is conducive to long-term, high-productivity work. Recognize that adding more light on the task may in fact aggravate the users' perceptions that the room generally appears dark. Typically this complaint is best addressed by reviewing room surface reflectances and/or surface luminances. *Assess the viewing conditions and tasks and ascertain the actual problem before recommending solutions. Recognize that employee education may be necessary to wean people from the incorrectly lighted environment to which they have become adapted.*

Luminaire. Commonly referred to as light fixtures, an assembly (usually consisting of a housing, lamp(s), reflector and/or refractor, wiring compartment, ballasts, and/or transformers) used to distribute light.

Luminance. The quantity of light reflecting from or transmitting through a surface in a specified direction of view *as detected by a meter.* Typically measured in candelas per square meter (cd/m^2) along with a specific direction or angle of "view." Absolute luminance of a given surface, area, or object has a significant influence on VDT veiling images and veiling reflections. The higher the surface luminance, the more likely that veiling images, veiling reflections, and/or glare are visible "in" or "from" the VDT screen—especially problematic with positive-contrast VDT screens. Luminance of a given surface, area, or object relative to another surface, area, or object is also important. If the surfaces, objects, or portions of surfaces or objects are relatively close in proximity, and the greater the luminance difference or ratio between them, then the greater the chance of these differences causing reflected images to appear "in" the VDT screen—again, especially problematic with positive-contrast VDT screens. If these surfaces, objects, or portions of surfaces or objects are relatively distant from one another in physical space, and the greater the luminance difference between them, then the greater the chance for transient adaptation effects and/or glare complaints and/or complaints of "high light levels" or "low light levels." *See also **brightness**, **luminance contrast**, and **luminance ratio**.*

Luminance Contrast. The difference in luminance between two or more adjacent markings, surfaces, areas, or objects. Luminance contrast is exhibited on this page— the luminance difference between the markings called "alphanumeric characters" or "letters" and the paper. Luminance contrast and **chromatic contrast** are responsible for how, what, and how quickly we see.

Luminance Ratio. Ratio between one area of brightness and another area of brightness. The greater the ratio, the more chance for glare, reflected glare, and veiling reflections in VDT screens.

Negative Contrast. Graphics (images) and/or text (alphanumeric characters) are darker than the background media. For example, the material in this book is presented as negative contrast, since the images and text, which are black and/or colored ink, are darker than the white paper (background media). Images and text on VDT screens can also be presented as negative contrast, although many of the earlier-generation VDTs (monochromatic—single-color green or amber) present material as positive contrast. Figure 1-6 shows a negative-contrast VDT screen. *See also **positive contrast** and Figure 1-5.*

Noninterlaced Monitors. The entire VDT screen is refreshed electronically from top to bottom, offering less annoying flicker. *Use only noninterlaced monitors with refresh rates of at least 72 Hz. See also **refresh rate** and **interlaced monitors**.*

Positive Contrast. Graphics or text are lighter than the background media. *See also **negative contrast** and Figures 1-5 and 1-6.*

Power Budget. Reported as a ratio of watts of power for all of the lighting to the area (square meters) covered by that lighting. Power budget provides some indication of energy efficiency. However, energy is the amount of power used over a period of time, so not only power but also the time of operation influence energy use. Hence, automated controls can be effective tools for reducing energy use. Lighting that meets user needs in the electronic office need not be energy intensive. Indeed, energy-intensive lighting in the electronic office feeds a vicious cycle of potential user discomfort by introducing heat that must be dissipated, extracted, or cooled.

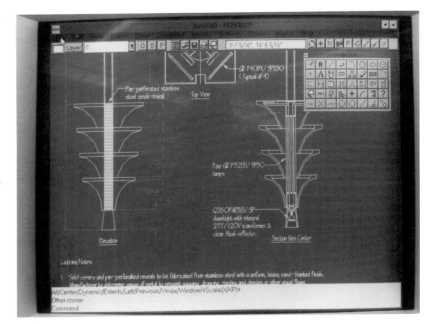

FIGURE 1-5

A positive-contrast display of information is shown on this VDT screen. Compare this with Figure 1-6, which illustrates a negative-contrast display. Courtesy of Fred Golden.

Refresh Rate. The rate at which the image/text on a VDT screen is refreshed electronically so that the image remains bright and visible. Higher refresh rates are best, preferably exceeding 72 Hz (72 times per second) to minimize screen flicker. Noninterlaced monitors offer the least flicker effect as compared to interlaced monitors. *See also **interlaced monitors** and **noninterlaced monitors***.

Retrofit Reflectors. Reflectors used in existing light fixtures that either have too many lamps producing too much light and using too much energy or have

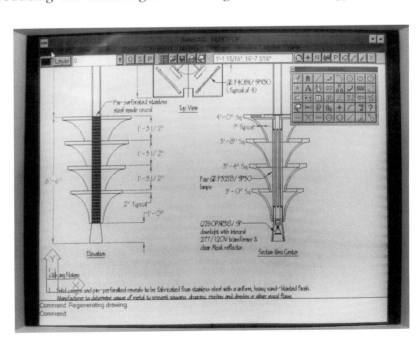

FIGURE 1-6

A negative-contrast VDT screen typically has a relatively bright background that mini-mizes veiling images and veiling reflections. Compare this screen set-up with that shown in Figure 1-5. Courtesy of Fred Golden.

degraded reflectors and produce too little light. These are not recommended for applications where all-new light fixtures are to be installed. Recognize that the retrofit reflectors may void any Canadian Standards Association (CSA), Underwriters' Laboratories (UL), or other standards-setting groups' ratings on existing fixtures. Recognize also that retrofit reflectors should always be photometered for the specific luminaire in which their use is being recommended. This is the only way to accurately assess optical performance. Consider a mock-up of a retrofit reflector to assess visual aesthetics and glare. Finally, recognize the physical limitations involved in lighting; claims by retrofit manufacturers that retrofit reflectors and half the number of lamps will yield similar or equal light levels as prior to retrofit and delamping simply are not true. If all conditions are equal, that is, the "old" lighting system is properly cleaned and relamped, and ballasts are in working order, then it is unlikely that inserting retrofit reflectors and taking out half of the lamps will yield light levels equal to those achieved under pre-retrofit conditions throughout the installation. Also, particularly relevant for VDT spaces, recognize the photometric implications of specular retrofit reflectors; if improperly designed and oriented, these will result in significant direct glare to the workers and veiling reflections and/or veiling images on the VDT screen.

Screenies. Aftermarket accessory for computer monitor housings that looks like a picture frame and acts like a bulletin board for mounting notes to the monitor housing. These should not be used. The posted notes act as visual clutter competing for visual attention with the VDT screen itself. Additionally, these surrounding "frames" may contrast with the VDT screen, particularly problematic with positive-contrast screens, resulting in visual fatigue and perhaps headaches. *Avoid the use of screenies.*

Surface Reflectances. The amount of light, reported in percentage, reflected from a surface versus the amount of light falling onto that surface. Surface reflectances are responsible for luminances and luminance ratios. See also *Surface Reflectances* in Chapter 3.

Transient Adaptation. Caused when an observer's view scans from bright conditions to dark conditions over a short period of time (e.g., looking from a dark VDT screen to a distant bright window or to nearby exceedingly bright paper documents). When viewing back and forth frequently from light to dark and vice versa (e.g., between a positive-contrast screen and an overlighted paper document), transient adaptation can cause visual fatigue. Maintaining appropriate luminance ratios and not exceeding absolute luminance limits can minimize this problem. *See also Task Surface Luminances* in Chapter 3.

Users or Observers. Tasks, lighting, and users interact to determine the success of any work environment involving visual work. Tasks and lighting can be well understood and well designed, yet if users are unfamiliar with tasks, lighting, or their own limitations, the result can be one of failure. Poor eyesight can only be partially remedied with improved task qualities and different lighting techniques. Users should have periodic eye exams, perhaps annually and certainly biennially, to ensure that corrective measures are taken to optimize vision.

Veiling Images. Image reflections that are rather sharply or crisply reflected from tasks. Even relatively dim objects, which usually are most pronounced on specular, positive-contrast VDT screens, can reflect clearly in the VDT screen. Two phenomena occur: First, the task image on the VDT screen is veiled by this object reflection, making it difficult to see the VDT task image; second, the worker's eyes periodically tend to focus on the reflected image. Since accommodation cannot simultaneously occur on objects of different focal lengths, the eyes accommodate between the VDT task image and the reflected image, eventually resulting in visual fatigue. Veiling images can be limited by using semispecular or matte VDT screens as well as using negative-contrast VDT screens and/or carefully following luminance criteria guidelines established in Chapter 3. See Figure 1-1.

Veiling Reflections. Also referred to as VDT screen washout, or "too much light" or "glare on the screen," these are reflections from a task that are significant enough to veil the text/graphic images of the task. This hazy glow or washout effect is not to be confused with veiling images. Veiling reflections are typically caused by high vertical illuminances on the screen and/or high luminances from surrounding surfaces or objects (e.g., windows, luminaires, bright surfaces). Maintaining appropriate vertical illuminances and luminance ratios and not exceeding absolute luminance limits can minimize this problem. *See also* discussions on specific criteria such as vertical illuminances, luminance ratios, and luminances in Chapter 3.

Video Display Terminal (VDT). A device that electronically displays graphic and/or text information on command (typically via a keyboard). Also known as a monitor or screen.

Visual Fatigue. Tired eyes, related to transient adaptation, accommodation, and glare—a relationship generally unknown to the observer but from which the effects are often experienced.

Basic Elements

Several basic elements of the electronic work place are responsible for the success or failure of lighting, that is, how well an observer can perform visual tasks. The architect, interior designer, electrical engineer, energy engineer, and/or lighting designer control some of these elements. The owner and/or facility manager control other elements. Still other elements are under the control of the observer. Finally, some of the elements cannot be controlled but can be understood as influencing the success of observers' performance of visual work. Table 1-1 outlines these basic elements and indicates where probable control over these elements may lay.

The table takes a simplistic view of these basic elements, which are essentially observer, task, and lighting related. Table 1-1 is only meant to indicate who has what kind of control or influence over the inherent characteristics of the basic elements that are ultimately responsible for how well people see, how comfortably people see, and how productive people are. For example, the computer screen hardware (the owner/manager controls the hardware used; the equipment manufacturer controls the quality and materials used to make the hardware) is inherently responsible for the ***potential visibility*** of the screen. Admittedly, adjustable work surfaces, VDT monitor stands, glare shields, and so forth may improve the screen visibility, yet these are ***aftermarket accessories***, none of which can ever increase screen visibility beyond its manufactured base condition. In fact, aftermarket accessories that can minimize veiling reflections and/or glare from VDT screens will affect character crispness and/or color discrimination and/or maximum character luminance. If the VDT screen is of poor quality and difficult to read, then it is likely more cost effective to purchase a better VDT screen than it is to purchase new lighting systems and new furniture.

> The computer screen, paper document, room surfaces, daylighting, electric lighting, and observers are the basic elements responsible for how well observers can perform visual tasks.

The remaining chapters of this book deal with those elements related to vision, and most specifically to lighting, over which the designers/engineers and users may have some control.

Computer Screen

> Important computer screen aspects include the following:
> ▼ Positive/negative contrast
> ▼ Refresh rates
> ▼ Image clarity or resolution
> ▼ Image size
> ▼ Screen finish
> ▼ Screen curvature
> ▼ Physical adjustment of housing
> ▼ Contrast adjustment
> ▼ Color vs. monochrome

Viewing the computer screen is one of the more important, if not the only, visual task performed in many of today's work environments. As such, more critical attention should be given to the selection of the computer screen. Initial cost of the screen should not be the sole driving factor in selection. Important aspects of a computer screen that should be reviewed include: positive/negative contrast; refresh rates; image clarity or resolution; image size; screen finish; screen curvature (e.g., convex, flat); physical adjustment of monitor or screen housing; contrast adjustment, color versus monochrome.

TABLE 1-1　　*Who Can Influence What and How: Basic Elements, Influential Participants, Probability of Influence, Level of Influence*

	Participants and Respective Influence			
Basic Elements	*Space Designers and Engineers*	*Owners and Managers*	*Workers*	*Equipment Manufacturers*
Computer Screen ▼ Monitor hardware ▼ Positive/negative display ▼ Screen maintenance	Unlikely ▼ Generally have little or no influence on computer hardware ▼ Should recognize that designers/engineers must accommodate hardware through design of furniture systems, lighting systems, room layouts, etc.	Likely ▼ Generally have final authority on computer hardware ▼ Should recognize implications of poor-quality hardware on productivity ▼ Should have capability to provide full-color monitors with maximum contrast options ▼ Should have capability to provide matte-screen monitors	Possible ▼ Should have capability to change to negative contrast—some monitors have a software "switch" to allow user/observer to change to negative contrast from positive contrast ▼ Should have capability to change screen contrast intensity via easily accessible contrast controls ▼ Can perform more frequent screen maintenance	Definitely ▼ Most anything is possible; it's a matter of money—how much a monitor manufacturer is willing to invest in R&D and/or in tooling; and how much owner/manager is willing to invest in hardware
Paper Document ▼ Color of paper ▼ Gloss of paper ▼ Inherent contrast of the medium (e.g., black vs. blue ink pen, felt-tip, pencil, dot-matrix printer vs. laser printer, etc.)	Unlikely	Possible ▼ Should have capability to change color of paper ▼ Should have capability to change gloss of paper ▼ Has capability to provide black-ink pens and felt-tip pens ▼ Has capability to provide and/or encourage proper printer ribbon maintenance ▼ Should have capability to provide paper-document stand	Possible ▼ Has capability to use black-ink pens and felt-tip pens ▼ Should report poor printer ribbons ▼ Should have capability to use paper-document stand to place paper at same distance/elevation from eyes as VDT screen	Yes ▼ Most anything is possible, it's a matter of money—how much is computer printer manufacturer willing to invest in R&D and/or tooling; what kind of paper quality/technology is used; and how much owner/manager is willing to invest in hardware

TABLE 1-1   (continued)

Basic Elements	Participants and Respective Influence			
	Space Designers and Engineers	Owners and Managers	Workers	Equipment Manufacturers
Room Surfaces ▼ Ceiling(s) ▼ Walls ▼ Floor(s) ▼ Worksurfaces	Yes ▼ Should forego design trends and/or egos in favor of more quantifiable/defensible decisions on finishes ▼ Should specify surface finishes that are of appropriate reflectances (see later discussions) ▼ Should specify surface finishes that are generally matte	Yes ▼ Should forego egos and image issues, or at least temper these in favor of more defensible decisions	Unlikely	Yes ▼ Ceiling, wall, floor covering, and furniture manufacturers can provide a palette of colors that have appropriate reflectances and are matte or satin in finish
Daylighting ▼ Atria ▼ Skylights ▼ Windows	Yes ▼ Should provide for large-area, low-to-moderate luminance, view-preserving media ▼ Should balance view-preserving media luminances with other room surface luminances ▼ Should develop universally appropriate shading techniques that allow all occupants a chance at view while simultaneously minimizing glare conditions for all occupants ▼ Should recognize the real reason for specifying "daylight" media—to allow workers a view to exterior world, not to generate thousands of lux of light for working	Yes ▼ Should forego lower initial-cost solutions in favor of longer-term payback in improved comfort and productivity ▼ Should forego long-held misbeliefs that low-transmittance glazing systems result in "dark and dingy" interiors ▼ Should be more open-minded to state-of-the-art and science applications	Possible ▼ Should have capability to individually control each or groups of window or skylight "coverings" (e.g., drapes or blinds)—recognizing that this seriously affects the working conditions of others	Yes ▼ Manufacturers of glazing and glazing treatments could provide more complete array of appropriate media that preserve color clarity while minimizing luminances

TABLE 1-1   (continued)

Basic Elements	Participants and Respective Influence			
	Space Designers and Engineers	Owners and Managers	Workers	Equipment Manufacturers
Electric Lighting ▼ Ambient ▼ Task ▼ Accent/fill ▼ Efficiency ▼ Subjective impressions	Yes ▼ Should forego design or hardware trends and/or egos in favor of more defensible decisions ▼ Should forego the attitude of simply addressing illumi-nances (light levels) and include such issues as lumi-nances and lumi-nance ratios (brightnesses) and subjective impres-sions ▼ Should forego the "…it's always been done this way …" attitude ▼ Should take a more active role in con-vincing or educating owners, managers, observers, and users of the correct way to do things	Yes ▼ Should forego egos in favor of more defensible decisions ▼ Should request more quantifiable data from design team to justify design team rec-ommendations ▼ Should be more open-minded to state-of-the-art and science applications ▼ Should recognize the extraordinary sums of monies paid in salaries and benefits and the subsequent positive influence of lighting on productivity	Possible ▼ Should have capability to control task lighting either in intensity (on/off or dim) and/or in orien-tation/location	Yes ▼ Most anything is possible; it's a mat-ter of money— what luminaire manufacturer is willing to invest in tooling and R&D; and what owner/manager is willing to invest in hardware (which is ultimately an investment in employees)
Observer ▼ Condition of eyes ▼ Preferences ▼ Nonvision issues (e.g., preconceived attitudes)	Possible ▼ Should account for the preferences of observers ▼ Should suggest to owners, managers, and/or observers the need for annual eye exams	Yes ▼ Should encourage design team to bet-ter understand and then account for observers' prefer-ences ▼ Should encourage observers to have annual eye exams ▼ Should encourage regular "task" breaks	Yes ▼ Should have annual eye exams ▼ Should express prefer-ences while recognizing limitations of democratic process and the limita-tions of good design practice, which tends to "meet most of the needs most of the time" ▼ Can take control of non-vision issues (e.g., revise job attitudes; develop a more positive outlook and a more involved role in work environment) ▼ Should take regular breaks from VDT screen-reading tasks	Unlikely

At the very minimum, purchasers of computer screens should always set up some short-term in situ program to test and assess proposed screens' capabilities. This may necessitate surveying a group of participants in a test program to judge the screens' successes or failures. Also, recognize that the in situ conditions should be as proposed if new or retrofit construction is to take place prior to the implementation of the new monitors. This may necessitate a mock-up of the proposed new or retrofit work environment in which to test the screens.

Screen contrast can minimize transient adaptation effects when viewing from paper documents to screen. Positive-contrast screens (light characters/graphics on dark backgrounds) generally have a much lower overall luminance than the reference paper task (which is assumed to have dark letters on white background—negative contrast). Negative-contrast screens (dark characters/graphics on light backgrounds) generally have similar luminance to the reference paper task. Figures 1-5 and 1-6 illustrate the differences between negative- and positive-contrast VDT screens. Figure 1-7 shows the use and relative brightness of a negative-contrast VDT screen in conjunction with a nearby reference paper document. This situation avoids significant contrast between the two tasks, paper and computer screen. Also, the lighter computer screen background helps to mask the light reflections that may eventually show up in the screen, depending on its final resting place and orientation within a work environment.[7,8]

Refresh rates can significantly affect the desirability of viewing a computer monitor. Most computer monitors use a phosphor coating that is excited on intervals by electricity. The frequency of these excitation intervals is known as the refresh rate. The more frequent the excitation intervals, the higher the refresh rate and the less

FIGURE I-7

This negative-contrast VDT screen has similar brightness characteristics as the nearby paper reference document. This minimizes, if not eliminates, transient adaptation effects. Courtesy Fred Golden.

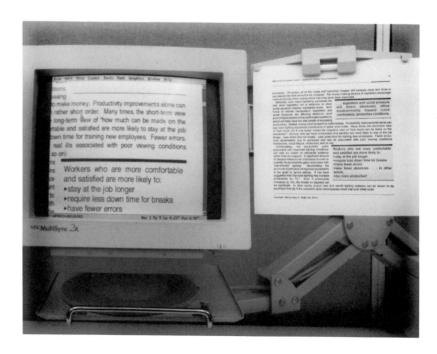

visible is flicker. Flicker is generally more visible and therefore more problematic on negative-contrast screens. Although a refresh rate of at least 60 Hz (cycles or number of times per second) is considered minimally acceptable for positive-contrast screens, a higher rate is necessary on negative-contrast screens.[9,10,11] Refresh rates approaching 90 Hz are preferable.[10,11,12] Also recognize that flicker is typically more bothersome in the periphery. Hence, when looking directly at the screen, flicker may not be noticeable; however, when turning the head to view nearby reference paper documents, flicker may be noticeable in the peripheral vision. As suggested previously, in situ testing of monitors can help discover such problems.

Refresh rates don't tell the whole story on flicker potential from a VDT monitor. Some VDT screens refresh every other line from top to bottom before returning to the top and refreshing the intermediate lines. These are known as interlaced monitors. Such monitors may have a very high published refresh rate, of say 90 Hz, but in fact this indicates that only every other line is refreshed 90 times a second, which results in an actual refresh appearance of 45 Hz. Forty-five Hz causes significant visual flicker, which can be quite bothersome to many people and result in headaches and visual fatigue after several hours of viewing. Monitors are available that refresh every line consecutively from top to bottom. These are known as non-interlaced monitors, and at 72 to 90 Hz will generally be acceptable for relatively long-term viewing.

Image clarity, also known as resolution, is important if the detection of small characters and/or graphics on the screen are expected to be part of the overall task. Obviously, blurred images, discontinuous lines, and/or graphics will slow the performance of work, if not increase the potential for error. Image clarity depends on both the VDT's dot-pitch (essentially how close together the electronic "dots" are—the closer together the better) and on internal graphics resolution hardware. A dot-pitch of 0.25 mm is desirable for extensive viewing of graphics. A larger dot-pitch can be acceptable for viewing text. Image clarity should be reviewed before purchasing the VDT monitor and the computer graphics hardware. Beware of reviewing image clarity characteristics in totally dark environments, unless the monitor screen is intended to be used in dark environments. Although images are clear in a darkened setting, in a lightened room the screen may require some sort of glare-reducing covering. After-market screen "filters" that are typically used to minimize screen glare and veiling images can degrade image clarity.

With any visual task, image size greatly affects the accuracy with which the task is performed. In general, the greater the image size, the greater the visibility. Image size on a VDT screen is typically a function of computer software. The monitor screen, however, must be able to accommodate increasing image size and maintain image clarity. This is based on the monitor's inherent image-producing capabilities and is best found when testing the monitor(s) in question with the intended software.

The screen finish has significant influence on the visibility of screen characters/graphics. Many full-color monitor screens have a specular, glossy finish to allow for the maximum contrast potential, both luminance and chromatic. Unfortunately, this maximum potential can only be realized in virtually blacked-out environments. Although the screen image is clear, few people are likely to desire working eight-hour shifts in black holes. Some full-color monitor screens have a semispecular finish that helps maximize contrast potential while reducing the few negative effects of even well-designed lighted environments on screen viewing.

Curvature of the screen interacts with the lighting conditions in an environment. Additionally, the screen finish may compound the visibility problem of more convex screens or enhance the visibility of more flat screens. For example, many of the less expensive, full-color monitors use glossy, standard, rather convex picture tubes—the front of the monitor is shiny and sort of bulges out from the corners of the screen toward the observer. Such glossy, convex screens actually do an excellent job of capturing the brightnesses on the ceiling and walls and reflecting these brightnesses directly into the observer's eyes. Figure 1-8 shows the effects of such a monitor. Flatter and more matte monitor screens are most desirable.

Physical adjustment (tilt and swivel) of the screen housing is desirable. New flat-screen technologies are arranged on a hingelike stand that allows the observer to adjust the viewing angle to meet the observer's own preferences.

On any VDT monitor, the contrast of the screen image should be easily controllable by the observer. Contrast knobs or wheels on the lower front or side are common and should be a necessity for those purchasing new monitors.

Based on the discussions on negative contrast, generally a color monitor is preferable to a monochromatic monitor. This typically not only enables the observer

FIGURE 1-8
The glossy, convex picture tube on this VDT monitor reflects all ceiling and wall brightnesses directly to the observer. Courtesy General Electric.

TABLE 1-2　　*Improving the Inherent Characteristics of Computer and Paper Tasks*

Basic Element	Parts	Attributes	Criteria
Computer	Screen	Best view	Tilt/Swivel
			500 mm to 700 mm from eyes
		Sharpness	0.25 mm dot-pitch
			15-inch or larger screen
			High-resolution graphics card
			Easily accessible contrast controls
		Contrast	Negative contrast
			Color
		Refresh rate	Noninterlaced
			72 Hz or greater
		Low glare	OEM nonreflective coating of screen
			Flat screen
	Software	Display setup	Light background
			Dark text
		WYSIWYG[a]	12-point or larger text
Paper Document	Paper	Finish	White or pastel
			Matte
		View	Document holder at same elevation and distance from eyes as computer screen
	Ink	Contrast	Matte black or saturated color

[a]WYSIWYG (pronounced "whiz-zee-wig") is computer slang for <u>w</u>hat <u>y</u>ou <u>s</u>ee <u>i</u>s <u>w</u>hat <u>y</u>ou <u>g</u>et and refers to a computer screen that displays the image exactly as it will be printed. This helps significantly in reviewing correct structure of correspondence, tables, and the like.

to set up a negative-contrast display but also to select background and text/image colors that enhance visibility and/or are most comfortable for the observer.

State-of-the-art monitors are available that have significantly improved optical viewing properties over earlier models and are less sensitive to the environmental conditions under which they are viewed. Table 1-2 summarizes the characteristics that should be specified for computer screens and paper documents. Most monitors in place now and likely to remain in place for another five to ten years or longer, however, *are* sensitive to the environmental setting. Lighting conditions become a significant factor in establishing observer comfort and productivity with these less-than-optimal monitors.

Screen and Monitor Maintenance

The human can adapt so well to many conditions that even as conditions deteriorate to "unacceptable," most people don't realize it is occurring or has occurred. Viewing conditions can be made quite poor simply by allowing dirt and fingerprints to build up on the monitor screen over time.

The monitor cabinet should also be maintained on a regular basis. For negative-contrast screens, the monitor cabinet should be of a medium-to-high reflectance. If paper notes clutter the monitor cabinet, these should be removed. These notes change the reflectance characteristics of the monitor case as well as introduce visual clutter that competes for view of the screen itself. "Screenies," the push-pin or enlarged note-mounting frames that can be placed around the monitor should be avoided—both to eliminate competing visual clutter and to maintain appropriate monitor cabinet reflectance.

Paper Documents

Data-entry VDT tasks usually involve reading both a VDT monitor *and* a paper reference document. These paper documents may range from handwritten pencil on yellow lined paper to laser-printed, black ink on white paper to blueline reproduction drawings of small details. Contrast for each of these various paper tasks varies greatly. Smaller-sized images, poorer quality images, and lower contrast images typically require rather significant levels of light for observers to be able to detect the appropriate detail and act on the detected information. Obviously this can present a significant challenge as the observer views from a paper document—which is lighted to very high levels to make the details visible—to a positive-contrast VDT screen that needs little if any light for the observer to easily review detail.

> Paper documents should be white or pastel background, matte finish, with crisp black or saturated color text and graphics.

Whatever can be done to improve the inherent readability of paper documents should be taken. White paper establishes an improved background for better contrast with any markings, images, or alphanumeric characters. Matte paper should be used, so that if/when localized or directional light is aimed onto the paper task there are few, if any, harsh glare reflections and/or veiling reflections—those reflections strong enough to actually veil the task, a classic problem with glossy magazines and now with VDT screens. Black ink and felt-tip pens offer better luminance contrast potential, with felt-tip pens generally preferable, since they leave a rather matte

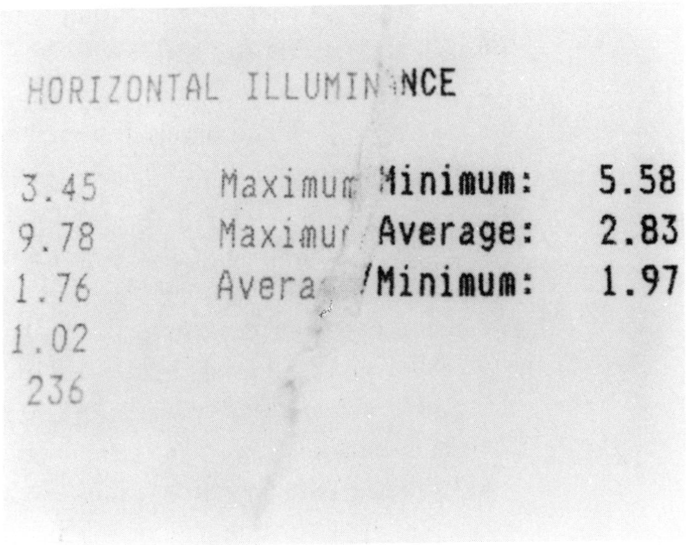

FIGURE 1-9
Same light conditions, but the text on left was produced prior to ribbon change (which was long overdue), whereas the text on right was produced immediately after ribbon change. Courtesy of Fred Golden.

marking on the paper, further minimizing glare reflections from the paper and minimizing veiling reflection problems.

If many paper reference documents are generated on computer printers, then a proactive maintenance program should be developed for ribbon and/or ink cartridge replacement. Figure 1-9 shows side-by-side identical documents generated on the same printer—one document generated with an old ribbon and the other with a new ribbon. The visibility difference is significant, with the document from the new ribbon easier to read, allowing for faster work with fewer errors. Additionally, if dot-matrix printers are used, consideration should be given to purchasing high-resolution printers, with a *draft* mode of rather high resolution (reviewing sample printouts before purchasing printers is suggested).

Paper-document holders that put the paper document at the same angle, elevation, and distance from the observer as the VDT screen are preferable options. This helps minimize visual fatigue by reducing accommodation as view changes from paper reference to VDT screen. Figures 1-4 and 1-7 illustrate the use of a paper-document holder.

Room Surfaces

Two aspects to surface finishes are important to consider—value and gloss. Value is an indication of the total amount of light reflected from the surface. Gloss is an indication of the *shininess* or directionality of the reflected light. Gloss is not indicative of *how much* light is reflected but only indicative of *how directional* the reflection of light is. For example, a black

> Value and gloss of room surface finishes affect how well people see VDT screens and affect how well people like a space.

glossy surface is likely to reflect only 2 percent of the light directed onto it. Nearly **all** of that 2 percent of light, however, is reflected in *one specific direction*. Figure 1-10 is a gray scale indicating the approximate reflectance values of various gray tones. Figure 1-11 graphically represents the three generic classifications of gloss; diffuse (or matte), semispecular (or semigloss), and specular (or gloss) relevant to room surface and VDT screen finishes. The VDT screen surface finishes influence how light is reflected from the VDT screen to the user, thus affecting veiling image and veiling reflection conditions. Room surface finishes have a marked influence on illuminances, luminances, luminance ratios, and subjective impressions. This means that room surface finishes influence how well people see VDT screens as well as how well people like a space. Room surface finish selection is one important step in establishing a successful electronic office. Just as carefully as the designer selects the lighting equipment, so too must the designer select room surface finishes. For example, the following simplistic approach to room surface finish selection for a room full of people using VDTs shows how solving just one part of the problem does not at all guarantee success.

If luminances and luminance ratios are excessive they cause reflections in VDT screens. Luminances and luminance ratios can be minimized if either or both of the following conditions are met: (1) illuminances on the "bright" surfaces are reduced; and/or (2) the reflectances of the "bright" surfaces are reduced. Therefore, if dark wall and ceiling finishes are selected (low reflectance values exemplified by dark gray or black), then these surfaces probably won't cause bothersome image reflections from the VDT screen. The room will look quite dark, however, and subjective impressions are likely to be negatively impacted unless quite a lot more light is added onto these dark-finished surfaces. This will involve the use of significantly more energy, however.

Room surface finishes also influence the directional intensity or severity of light reflections. In the preceding example, if the darker wall and ceiling finishes are

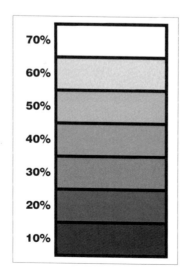

FIGURE 1-10

Gray scale with respective reflectances.

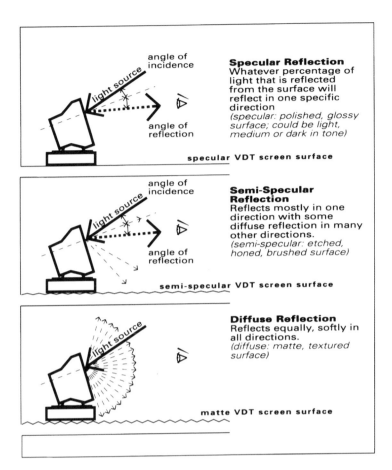

FIGURE 1-11

Three generic VDT screen and room surface gloss classifications.

specular or shiny, then depending on the location and orientation of VDT screens, enough light may reflect in one specific direction to cause reflected images to occur in one or more VDT screens. A seemingly simple decision of room surface finishes has significant, multiple ramifications.

Daylighting

Daylighting has historic, traditional, and romantic connotations to many people. Indeed, historically, given the kinds of visual tasks performed and the lack of artificially generated light, except in the form of candle, whale oil, kerosene, or gas light, daylight could be introduced "full strength" through whatever apertures looked best from inside and/or outside as well as whatever apertures permitted daylight entry to most significantly enhance the religious, work, or living experience.

Unfortunately, for those people using computer screens as a means of performing work, daylighting introduced in the historic or traditional methods will certainly exacerbate if not simply introduce poor visibility of computer tasks.

Why should daylight media be used at all in today's electronic workplaces? Daylighting should be considered for illuminance, for luminance balancing and luminance patterning, and, perhaps most significant, for the visible connection to the exterior environment. Using daylighting for the issues of illuminance, luminance balancing, and luminance patterning can yield significant energy savings. Using daylighting for the issue of visible connection to the exterior environment can yield significant users' satisfaction and, hopefully, productivity gains.

Avoiding problems with daylighting in the electronic workplace, however, requires design solutions that recognize and respond to the design issues associated with the performance of computer tasks. For example, skylights should be designed to provide very large, continuous daylight/view media. The image-preserving skylight is not only a means of introducing daylight onto the task area and luminance onto the "ceiling" plane but also a means of providing workers a view to the exterior—even if only a view of sky. The worker has a better sense of time; has the benefit of changes of cloud views and sky color; and a significantly better connection with the exterior than with a nonimage-preserving medium. A large-area skylight can provide a monolithic plane of luminance. The key is to develop a skylight system that limits maximum luminance and maximum illuminance. Maximum illuminance issues are related to direct solar radiation—the sun shining directly onto a skylight will generate the maximum illuminance on the skylight and subsequently the maximum light coming through the skylight and falling onto tasks and task areas. Typically, maximum illuminance on skylights occurs about noon in the summer months in the Northern Hemisphere, when as much as 100,000 lux may fall on the skylight. More average conditions may result in 50,000 lux on the skylight. If paper and VDT tasks are not to be washed out when located under skylights, then skylight transmittances must be quite low. Transmittances of between 2 percent and 10 percent are not unreasonable. Indeed, transmittances greater than these values are likely to lead to serious environmental deficiencies and, subsequently, occupant

FIGURE 1-12

If VDTs must be used in an area adjacent to windows, the VDT screen should be oriented so that the observer's line of sight is parallel with the window wall. This limits the direct glare and veiling image effects caused by the high luminances of the window wall.

complaints and lost productivity. General guidelines on VDT and observer orientation like those demonstrated in Figure 1-12 can help minimize the impact of daylight problems. Nonetheless, planning flexibility is compromised. Therefore, glazing treatment is a key element in appropriately designing daylighting and electric lighting for electronic work places.

Figures 1-13 and 1-14 provide some direct insight on glazing treatment possibilities and their resulting effects. Figure 1-13 illustrates a mock-up of two glazing treatments. Figure 1-14 shows the veiling reflections present on a VDT screen (with no text) as a result of the daylight transmitted through the two glazing treatments shown in Figure 1-13. The partly cloudy conditions of a late fall morning near Pittsburgh, Pennsylvania, USA, provided a good test, as luminances of hazy and partly overcast skies tend to be quite high. Imagine the resulting veiling reflections if glazing transmittance exceeded 13 percent—which is the case in many situations because little attention is paid to daylighting issues.

Limiting luminances is just as problematic as limiting illuminances. For example, in the northeastern, midwestern, and northwestern United States, skylights could be designed for luminances caused by partly cloudy, summer-sky conditions. For skylights on buildings that are not shadowed by other buildings or obstructions, this might be a "worst-case scenario." Yet depending on the likelihood of this worst-case scenario happening, and/or depending on the severity or arduousness of the work expected to be performed under the skylight, it might be too conservative (and also too costly) to base glazing design and skylight shading treatments on this worst-case scenario.

Of course, one of the reasons a designer (architect, interior designer, lighting designer, and/or electrical engineer) is retained is to help the client sort out and establish the design conditions to which the building and its systems should be designed. This likely means developing a list of goals or achievements for the building and then ranking these anticipated achievements from "absolutely necessary" to "minimally necessary" to "desirable." For example, if VDT tasks will be prevalent, and occupants will perform VDT tasks during most of the day, then it is arguably "absolutely necessary" to limit skylight luminance based on an anticipated worst-case condition—that being a hazy or partly cloudy day, midsummer, at about noon when sky luminances could intermittently exceed 18,000 cd/m^2. This would create a potential glare problem for the occupants. Alternatively, if VDT tasks will be performed infrequently and/or for short durations, it is arguably "desirable" to minimize skylight luminance for many of the days throughout the year, but not necessarily for the worst day. Given the likely percentage of overcast versus clear conditions, the overcast condition could arguably be used as the basis for the design model, in which case sky luminance would be perhaps 10,000 cd/m^2. Hence, this skylight glazing system could be of higher transmittance than the skylight glazing system used for the 18,000 cd/m^2 condition. In fact, the skylight glazing system for the 10,000 cd/m^2 could be nearly twice the transmittance (18,000 ÷ 10,000).

FIGURE 1-13

*A mock-up study of several glazing treat-
ment shows two windows each with different tints and frit densities and cover-
ages (frit is a ceramic coating which can
be opaque or translucent). The partial frit
coating helps minimize sky and solar disc
(sun) luminances. The window on the left
has a base transmittance of 13 percent,
while the window on the right has a base
transmittance of 8 percent. Both windows
have a partial frit pattern that further
reduces transmittance without spectrally
altering the view (contrary to the effect of
additional tints and reflective coatings).
See Figure 1-14. Courtesy of Gary Steffy
Lighting Design Inc.*

FIGURE 1-14

*Figure 1-13 is a view of a
glazing mock-up from
inside the mock-up space.
Here, a close-up of a VDT
screen (with no text or
graphics) shows the lumi-
nance effects of two dif-
ferent types of glazing
treatments under consid-
eration. The 8 percent
glazing with partial frit
(veiling reflection on left)
is less obtrusive than the
13 percent glazing with
partial frit (veiling reflection on right). Courtesy of
Gary Steffy Lighting
Design Inc.*

At the same time, skylight design must account for illuminances that may be present. As indicated previously, worst-case conditions are likely to be clear, summer days, around noon, when illuminances generated by the sun can reach 100,000 lux. Designing skylight transmittances low enough to minimize the severity of 100,000 lux is likely to result in transmittances low enough to address the luminance conditions. Both illuminance and luminance issues must be checked, however, to ensure that an appropriate transmittance is selected for skylight glazing.

On the other hand, window walls for the northeastern, midwestern, and northwestern locations arguably should be designed for partly cloudy, winter-sky, snow-on-the ground conditions. Window glazing systems' design will depend, just as it does for the skylight scenarios discussed, on the anticipated tasks, task duration, and/or task frequency.

Glazing systems for buildings with occupants expected to work on VDT tasks should be low in transmittance to meet luminance and luminance ratio criteria. Because of commonly held "design philosophies," many glazing systems of the past that are inappropriate for electronic office workers are still designed today on many buildings—new and retrofit. Glazing system transmittances need to be very low. For physiological (health) reasons, a good dose of ultraviolet and visible radiation for half an hour to several hours or so a day seems appropriate. Literature attests that for visible radiation to have an impact on the physiology of humans, it is likely that intensities of light (lux levels) need be high (typically greater than several thousand lux) but may only need to occur for several hours or less in duration.[13,14] Nevertheless, it is counterproductive to attempt to introduce such high doses of daylight or electric light in the electronic office throughout the workday to meet the physiological needs of humans, needs that could be addressed in "health" rooms at break times or with "bursts" of light at break times within the work setting. High-level doses should be achieved through daytime outdoor breaks, break areas that have high levels of daylighting or electric lighting not in the proximity of electronic office areas (e.g., lobbies, lounges, atria that are at some distance from the office areas), or educating workers to seek daylight exposure pre- or postwork hours. Indeed, there is cause for concern if significant ultraviolet levels are sustained throughout the workday in the work environment.

A revolution needs to take place in daylighting design in buildings housing people using computers.

▼ Low transmission image-preserving glazing systems are not only acceptable but necessary.

▼ Low transmission is achievable in a variety of ways, including techniques inherent to glazing and/or automated window-shading devices.

▼ Recognition that ordinary horizontal and/or vertical blinds do not meet any of the reasons for installing glazing in the first place—they block both view and light.

Daylighting design needs to be revolutionary, with

▼ luminance- and illuminance-driven designs rather than appearance-driven designs,

▼ low-transmission image-preserving glazing.

▼ low-transmission image-preserving window treatments, if any,

▼ automated control of window treatments and electric lighting,

▼ stopping the design- and cost-intensive use of so-called passive and active high-illuminance-level daylight collector systems, and

▼ stopping the use of antiquated horizontal and vertical blind window treatments.

▼ The control of daylight must be viewed similarly to the control of heating/ ventilating/air conditioning—it must be automated to activate at predetermined design conditions to meet the visual needs of most everyone at most times in the electronic office.

▼ More attention needs to be directed toward selecting glazing systems offering relatively accurate views (e.g., no dark-gray or deep-blue reflective coatings or tint, which make all views look discouragingly heavy-overcast).

▼ Less effort needs to be directed toward "active" and "passive" daylight collector systems.

▼ The money saved on the active and passive daylight collectors must be used to purchase dimmable, photocell-controlled electric ambient lighting systems.

▼ Energy savings are still substantial even with these reduced daylight levels (later discussions in both Chapter 2, *Worldwide Lighting Influences*, and Chapter 3, *A Model Lighting Guideline*, will show that we are not attempting to generate 750 to 1500 lux of general lighting from daylight in the VDT-intensive workplace but that general or ambient light levels need only be 300 lux or so; hence, lower daylight levels can still reduce energy requirements significantly if thermal properties are appropriately addressed).

▼ Energy savings are likely to be realized more frequently since people won't be closing opaque blinds to eliminate the harsh washout and glare from higher-transmission glazing systems—shutting blinds or drapes defeats the purposes of view and so-called free daylight.

Electric Lighting

Why is it that lighting is even an issue, let alone an important issue? Lighting should be used to help people do the following:

▼ See well enough to satisfactorily perform the visual parts of work tasks

▼ Be visually comfortable within the work environment to sustain the duration of work over time (typically an eight-hour period)

▼ Use energy resources most effectively

The difference between the first two reasons for using appropriate lighting is significant. To see well enough to be productive and satisfactorily perform the visual parts of work tasks, the typical, disease-free pair of eyes only needs 100 to 215 lux to adequately detect printed or handwritten information. With many electronic tasks, literally no external light is required for a pair of eyes to see the information displayed; in fact, external lighting can "mess up" the computer display task, reducing one's ability to adequately see it. Yet many people aren't comfortable working in a "blacked-out" room. Additionally, focusing on electronic tasks with all surrounding surfaces unlighted can cause as much visual fatigue and contribute to as many headaches as having surrounding surfaces lighted too much. Therefore, lighting must be addressed for at least two reasons—to see well enough and to be visually comfortable (recognizing other kinds of comfort, including thermal comfort, aural comfort, "ergonomic" comfort with chairs and work surfaces, and so on).

The third reason—to use energy resources most effectively—is typically sacrificed for lower initial costs, even though life-cycle costs generally favor energy-prudent designs. Additionally, the third reason is included here because many times it is not fully exploited in the realm of lighting as a double-edge sword. Admittedly, lighting can be designed to achieve a low connected load (11 watts or so per square meter). If lighting energy is curtailed at the expense of productivity, then no energy is saved—it simply takes more time (lights are energized for longer periods, hence use more energy than originally presumed) or takes more people (lights need to be used over a larger area that is necessary to accommodate more people, hence more energy used for lighting). Lighting can and must be designed not only to achieve low connected loads but also to maintain **or improve** productivity. Although the cost of productivity, or "producing goods and services," accounts for the biggest portion of corporate expenses (employees' salaries and benefits), lighting is hardly ever considered as having an impact—positively or negatively—on it. What a shame that such a small portion of corporate expenses (lighting) is not recognized for having such a significant impact on productivity and hence on corporate expenses.

Lighting for electronic offices can be categorized in one of three ways: ambient, task, and accent/fill. *Ambient lighting* refers to either the effect or equipment responsible for the effect of general lighting throughout the workspace. *Task lighting* refers to either the effect or equipment responsible for the effect of specifically lighting task areas (work surfaces or

> There is no hardware panacea for lighting electronic offices.

desks and reference areas and paper-document stands) throughout the workspace. *Accent/fill* lighting refers to either the effect or equipment responsible for the effect of accent or fill lighting of specific architectural features or areas, which is necessary to augment ambient and/or task lighting for the proper luminance balancing or luminance patterning throughout the workspace. None of these lighting techniques/categories is frivolous to the designer using defensible objective and subjective lighting criteria. With careful forethought given to criteria, criteria priorities, and lighting solutions, ambient, task, and accent/fill lighting can be achieved while meeting efficiency (for everyone's benefit), productivity (for the employer's benefit), visibility (for the employee's benefit), and comfort (for the employee's benefit.

Ambient lighting can be achieved with just about any hardware and lighting technique, including, of course, daylighting, but this particular discussion focuses on electric lighting. Indeed, there is no hardware panacea for resolving the problem of ambient lighting—the panacea is the criteria! Once criteria are defined, design resolution and equipment selection can be straightforward. Direct, direct/indirect, indirect/direct, and indirect lighting equipment can all be used successfully in addressing ambient lighting in the electronic office. Figure 1-15 shows the various generic equipment distributions.

▼ Direct lighting is that equipment which directs the light outward and downward, not putting any light above the horizontal plane defining the bottom of the luminaire.

▼ Direct/indirect lighting is a reference to equipment that directs most of the light outward and downward but with perhaps as much as 20 percent of the light going upward.

▼ Indirect/direct lighting is that equipment which directs about 80 percent of the light upward and perhaps as much as 20 percent outward and downward.

▼ Indirect lighting equipment directs all of the light upward.

Parabolic louvered luminaires aren't the solution, unless very specific luminaire luminance criteria are used and providing luminance balancing and patterning within the space is addressed. Indirect luminaires are not necessarily more energy intensive than direct luminaires—especially when one recognizes that direct luminaires using the most efficient lamps will require some sort of significant shielding to prevent glare and washout on the VDT screen. Luminaire shielding is typically

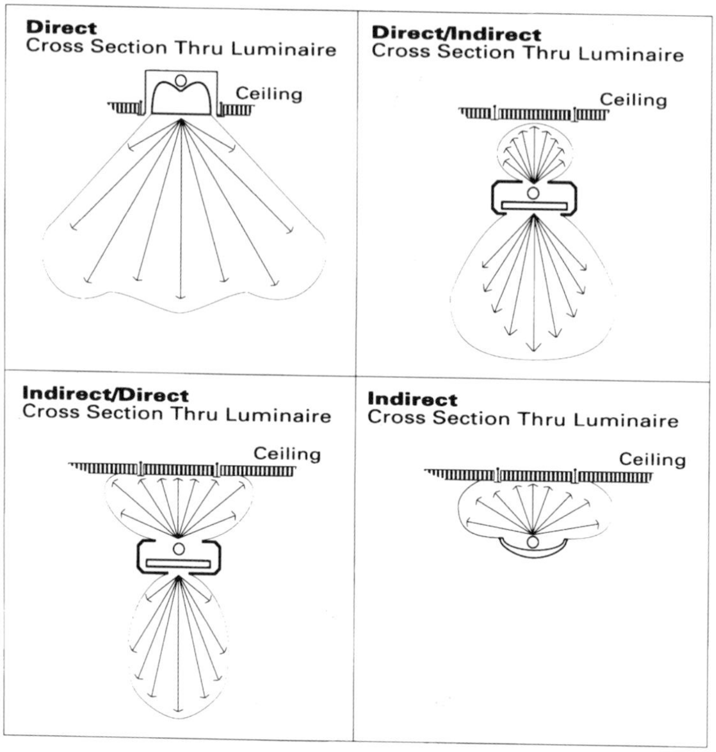

FIGURE 1-15

The four generic lighting approaches to ambient lighting, all of which can succeed for people using computers if lighting criteria are enforced during design.

achieved at the expense of efficiency. In other words, for direct or direct/indirect lighting equipment, there is a trade-off between glare and efficiency. Of course, indirect luminaires have their drawbacks, too—mounting heights, introducing the appearance of secondary ceiling planes, luminaire luminance versus ceiling luminance, and so on.

Task lighting, while offering the potential of great user-interaction or "high-touch," is many times seen as a decorative element. Equipment is selected for its style or its price. Rarely is task lighting selected for its functionality or its ability to allow the worker to fine-tune or adjust it to his or her liking. Lighting equipment should not be selected as design fashion. Compact, sleek halogen task lights are heat sources, excessively glary, energy intensive, and high-maintenance shadow producers.

Accent/fill lighting is typically overlooked—because it is perceived as an unnecessary expense. For several key reasons, accent/fill lighting is critical to the success of electronic workspaces. First, even if ambient lighting and task lighting are properly designed, there are likely to be zones or areas where, because of the space layout, workstation layout, circulation requirements, wall surface finish changes, ceiling finish surface changes, etcetera, luminance balancing and patterning is subpar, that is, below criteria. To meet criteria, some light needs to "fill" these voids. Second, for visually distant focuses to be used for eye-muscle relaxation (see *Accommodation* under the earlier *Key Concepts* section), some definition to those distant focuses is necessary. For example, a view outside offers the worker who reads the computer screen for long periods of time a chance to glance away now and again for visual relief. The significance of the view and its relative brightness (even with the low transmittance glazing discussed previously in *Daylighting*) suffices to attract attention. If exterior views don't exist, or if some people are far from windows, then interior distant focuses become important. Poster art, wall surface color and texture patterns of rather bold style can serve this purpose of distant focus, as can any sort of artwork, foliage, and even signage. Lighting these distant focuses, however, is generally necessary to provide the visual attention and aid in focus identification by the distant observer. This accent lighting is as much a part of the overall lighting scheme as are task and ambient lighting.

Observer

Certainly, the best-laid plans can still go wrong. The "observer" or worker holds the potential to do "good work"—as defined by the employer. Poor lighting will debilitate those workers with the potential to do good work and the willingness to use that potential. Good lighting will allow those workers with the potential and willingness to use it to excel. Hence, good lighting needs to be delivered to the workplace if there is to be success in the work-

> Poor lighting will debilitate those workers who have the potential to do good work.

place. Additionally, however, the worker needs to be "maintained" periodically to ensure that disease and/or age effects are kept in check (e.g., with eyewear) or accommodated in design. This may entail encouraged eye exams and physical exams on a periodic basis.

Worker preferences, which may vary significantly from worker to worker should be considered when developing lighting criteria. For example, if one worker prefers higher lighting levels on paper tasks than another worker, then task lighting can be designed/selected to accommodate this difference in preference—either by using more task lights, more lamps in task lights, or dimmable task lights for best flexibility as workers move around.

Workers' knowledge on VDTs, visual work, and work habits may need to be addressed to minimize negative attitudes and/or preconceived notions on what will and will not work. An example discussed previously is the attitude of viewing VDTs in the dark. This is a prevalent attitude. Given the choice, many people are likely to want a "darkroom" condition for working on VDTs not realizing the psychological implications of working in a black hole. People need to be taught that electronic workplaces can be lighted quite satisfactorily if appropriate criteria are maintained and providing the solutions are properly maintained. Many workers don't realize the significance of simply maintaining the VDT screen. Daily cleaning of the screen can enhance its visibility (clarity) or can ruin the glare-reducing coating on it. All workers should know how to properly maintain the VDT screen or understand that periodically some other individual will be cleaning the VDT screen. Additionally, workers need to know that breaks from VDT work are encouraged on a periodic basis. Further, they should be instructed on the benefits of visually—distant focusing (accommodation) on a periodic basis.

A descriptive booklet and/or short course or video (perhaps half an hour to an hour) on VDTs, visual work and work habits, and lighting's role should be considered a part of corporate training programs.

What's It All Mean?

People using computers can be more comfortable and more productive. Perhaps more time designing the lighted environment is required. Perhaps making more careful selections of computer hardware and software is required. Table 1-2 illustrates some of the computer and paper characteristics that can help people see tasks better. Perhaps being more cognizant of one's own health is required. Making decisions based on historic, romantic images of design, on first cost only, or on popular-press sound bites is just plain wrong.

> Making decisions based on historic, romantic images of design; or on first-cost only; or on newsy "revelations" is just plain wrong.

Historic, romantic design applications (e.g., large orifices in many Gothic cathedrals allow light to "stream in") are likely to lead to indefensible and inappropriate solutions for the electronic office workers. Such daylight "streaming" will wreak havoc with VDTs and workers. Making decisions based solely on initial costs regardless of life-cycle costs is likely to lead to significant, uncorrectable (without throwing out the new, inferior stuff) work situations. Following the advice of popular-press sound bites regarding health (e.g., lots of full-spectrum lighting will make us healthier) may actually lead to other health issues in the workplace. Full-spectrum lighting is a misnomer, and too much of this so-called full-spectrum light (or any light) can lead to glare, veiling reflections, and even risk of overexposure to ultraviolet and/or infrared radiation.

Consideration should be given to following a design process that reviews and addresses quantifiable and defensible lighting issues. Lighting guidelines must be a part of the design process.

2 | Worldwide Lighting Influences

Electronic tasks have introduced an entirely new and different set of viewing conditions for people. Specifically, the electronic task, for the most part, is internally illuminated and glass faced—both elements being quite sensitive to ambient room lighting. Further, the paper tasks used as reference to the electronic task require fair levels of light to be easily discerned. Finally, the majority of VDTs are now in spaces designed prior to the VDT's introduction as a visual task, necessitating retrofit of existing lighting systems originally meant to light paper tasks exclusively.

A common goal in the design of any work environment should be achieving the potential for optimum productivity—providing an environment that maximizes visibility of all expected visual tasks and maintains appropriate levels of visual and other physical comforts. Manifested by worker concerns about comfort and health, a number of guidelines have been drafted worldwide over the past decade. In some instances, these drafts have led to recommended voluntary guidelines. In other cases, these drafts have been legislated as standards. In still other instances, these drafts have remained as drafts, since the electronic office continues to change at a rapid pace. These guidelines, standards, and drafts all purport to address the many environmental issues in electronic offices. Lighting systems that were once state-of-the-art and quite appropriate for most hard-copy visual tasks performed are not suited for VDT-intensive office environments. Implementation of lighting conditions suitable for electronic offices is suspected to result in reduced employee complaints and improved working conditions and may lead to increases in productivity.

> A common goal should be providing an environment for workers which maximizes visibility of all expected visual tasks while maintaining comfort.

A review of the literature on lighting guidelines for electronic offices reveals a wealth of information. Of particular importance is the number of guidelines on related issues, such as furnishings, furniture arrangement, workstation configuration, etcetera, that are available. Many of these references have extensive recommendations on lighting issues. The designer/engineer is encouraged to consult these references for information on related issues as well as additional information on lighting. Table 2-1 outlines several exemplary guidelines representing a variety of views on the topic. The references listed in Table 2-1 are likely to be available for a charge through the contacts listed. To keep abreast with changes in technologies, these references are likely to change every five to ten years.

There are many other guidelines and standards regarding the design of electronic offices. Those listed in Table 2-1 soon may be, if not already, revised to account for changing technologies—in furnishings, computers, lighting, etcetera—or to account for changes in practice or perceived appropriateness of previous practice. Some guidelines now in print may cover areas other than lighting quite well but fail to grasp the significance of lighting or simply fail to understand lighting. Particularly disappointing with respect to lighting is the 1988 ANSI *American National Standard for Human Factors Engineering of Visual Display Terminal Workstations.*[15] Citing only a few then-recent studies, the ANSI/Human Factors Society document does not recognize the significance of or properly address the lighting-related human factors in electronic workplaces—luminances and luminance ratios—that have been shown empirically to have significant influence on VDT-reflected images, veiling reflections, and glare. These luminance issues *are* addressed directly, however, by many global guidelines on electronic office lighting. Luminance ratio issues are addressed directly by a select few of the global guidelines, whereas others address these issues indirectly with reflectance and illuminance criteria.

To stay abreast of current practice, research, guidelines, and standards, there are two reference resources to consult in addition to those listed in Table 2-1:

▼ Global Engineering Documents (provides a document search and sales service)
Division of Information Handling Services Inc.
2805 McGaw Avenue
Irvine, California 92714

▼ *VDT News/The VDT Health and Safety Report* (provides a subscription service)
P.O. Box 1799
Grand Central Station
New York, New York 10163

The German standard referenced in Table 2-1 is quite rigorous. All issues with respect to glare and veiling reflections are well covered. Since the DIN lighting standard is one part of a standard that is itself part of a series of standards, there is quite

TABLE 2-1   *Selected Worldwide Guidelines and Standards*

Document	Sponsoring Organization	Contact/Address	Comments
Artificial Lighting of Interiors: Lighting of Rooms with VDU Workstations or VDU Assisted Workplaces[16] [DIN-1988]	*Deutsches Institut für Normung*	*American National Standards Institute 11 West 42nd Street New York, NY 10036 USA*	*Somewhat tedious reading, but this is Germany's federal law.*
Guidelines for the Use and Functioning of Video Display Terminals, Part 1[17] [NJ-1989]	*New Jersey Department of Health*	*New Jersey Department of Health CN360, Room 701 Trenton, NJ 08625–0360 USA*	*Provides a glimpse of potentially legislated approaches in the United States. Covers a wide range of topics and issues. Lighting section covers basic principles but lacks objective (quantifiable) depth.*
IES Recommended Practices for Lighting Offices Containing Computer Visual Display Terminals[18] [IESNA—1990]	*Illuminating Engineering Society of North America*	*Illuminating Engineering Society of North America 120 Wall Street 17th Floor New York, NY 10005 USA*	*Comprehensive review of lighting issues. Cites current practice approaches, useful illustrations.*
Lighting Guide: Areas for Visual Display Terminals[19] [CIBSE—1989]	*The Chartered Institution of Building Services Engineers (CIBSE)*	*The Chartered Institution of Building Services Engineers Delta House 222 Balham High Road London SW12 9BS UK*	*An excellent application-oriented treatise.*
Vision and the Visual Display Unit Work Station[20] [CIE—1984]	*International Commission on Illumination*	*International Commission on Illumination/United States National Committee c/o ARC Sales 7 Pond Street Salem, MA 01970–4893 USA*	*Definitely has useful information. Extraordinary detail in some areas, especially those not typically directly under the control of the designer (e.g., VDT screen luminances and contrast), yet not expanding this to the environmental factors typically under the control of the designer. Many generalities and few specifics.*

a bit of criteria cross-referencing. These standards compose the German federal law with respect to workplace design. To meet these standards, highly refined lighting equipment is typically required. Environmental settings need to be relatively straightforward and spartan to meet the reflectance, luminance, and illuminance requirements set forth. As might be expected, the German DIN standards require well-engineered solutions. The DIN standard lighting section cited in Table 2-1, and later in Table 2-4, does not explicitly address the subjective issues of observer preference or spatial pleasantness. Criteria are reported in metric (SI) units.

The New Jersey guidelines were developed and implemented for environments housing state employees. Although these guidelines cover a range of issues related to the electronic office, the lighting segment is rather anecdotal and lacks in citing objective criteria to support many of the recommendations. This is not to say the anecdotal segments are useless; indeed, as of their issue date in 1989 and through 1992, these recommendations are helpful. Nevertheless, as technologies change and new lighting products become available, the designer will have little to reference in determining the new technologies' appropriateness and/or proper application. Sweeping generalities such as "indirect room lighting is preferred," although well intended, ill-serve the design community. Indirect lighting can create ceiling reflections more intense than the brightness of some direct lighting solutions. Criteria are reported in metric (SI) units with English conversion.

Although not introduced until 1990, the Illuminating Engineering Society of North America (IESNA) guidelines were in the works for seven years. This gave the organization the advantage of reviewing the evolution of other guidelines and standards on lighting for the electronic office. Additionally this enabled the IESNA to have significant practitioner review, "experimentation with proposed recommendations," and input. The result is a comprehensive document on the lighting-related issues of electronic office design. Perhaps the single disappointment with such a document is the lack of exemplary illustrations rather than generic drawings. This appears to be the single compromise, however, from such a diverse committee. Criteria are reported in metric (SI) units (with English conversion table at end of document).

The Chartered Institution of Building Services Engineers (CIBSE) was perhaps one of the first organizations to introduce guidelines for electronic offices with its *Technical Memoranda TM6: Lighting for Visual Display Units*[21] published in 1981. Between that time and 1989, the issue date of its *Lighting Guide* cited in Tables 2-1 and 2-4, the CIBSE, like the IESNA, learned a great deal in the history of application of earlier recommendations and technical limitations of available equipment. The later *Lighting Guide* is easy to read, yet contains objective criteria useful to any designer in any electronic office situation using any manufacturer's lighting equipment. Additionally, the *Lighting Guide* is well illustrated. Criteria are reported in metric (SI) units.

The International Commission on Illumination released its guidelines in 1984. As such, the document requires revision to update application technologies and criteria. Where definitive information was available in 1984, it is included in this document. Yet a fair amount of this material is of little direct use to the designer, since little discussion is included. General recommendations are made where research and/or application data were lacking at publication date. In some instances this results in rather large ranges of values cited as appropriate. Criteria are reported in metric (SI) units.

Lighting issues are addressed in the five guidelines/standards cited in Table 2-1. Other issues, including furniture, VDT monitors, support accessories, and the like are also addressed. The guidelines/standards cited in Table 2-1 have some common lighting issues and criteria of particular interest. A review of various other worldwide references, including several draft documents, also reveals

---

Lighting is not an exact science and is partly an art. Lighting criteria are norms, averages, or ranges that satisfy most of the people most of the time. Lighting criteria are design targets.

common lighting issues and criteria. An overview of such lighting-related issues as illuminances, luminances, luminance ratios, surface reflectances, and some miscellaneous criteria (which vary from guideline to guideline) is outlined in Table 2-4. This overview of criteria and the detailed discussion of relevant lighting criteria that follows are used as a basis for developing *A Model Lighting Guideline*, presented in Chapter 3.

Lighting Issues

Objective lighting criteria for most any environment, including electronic offices, can probably be codified by illuminances, luminances, and luminance ratios. Certainly there are other lighting issues that should be addressed for most any space or area where people are expected. For example, subjective issues of pleasantness and preference should be considered. Issues such as energy consumption must be considered. All of these issues will be addressed throughout this text, most prominently in the next chapter. In reviewing worldwide guidelines on lighting for electronic offices, however, the objective issues of illuminances, luminances, and luminance ratios are recurring. These issues will be briefly addressed here to establish their necessity.

Every person is a "light meter." Every person is therefore a lighting designer. As such, there are few, if any absolutes in lighting. What "just works" for one individual may be "perfect" for another and may be "disappointing" for yet another. Humans are particularly adaptive. We can live and work under a wide range of conditions. Unfortunately, we do not have a very good metric or measurement for determining the toll that is taken for living and working over this range. Hence, we attempt to establish norms or averages that satisfy most of the people most of the time. For these reasons, lighting is not an exact science. It is still partly an art. Nevertheless, we should not ignore the potential comfort and productivity gains that may be forthcoming if we design our environments within some set of conditions that are based in research and/or practice. Henceforth, all lighting criteria should be seen as design targets.

Illuminance

Illuminance is the amount of light falling onto a surface, object, or area. For electronic workplaces, illuminance is a double-edged sword. For data-entry tasks, where paper reference documents are prevalent, a sufficient quantity of light is necessary on the paper document if the observer is to successfully and comfortably read the paper document. At the same time, however, the VDT screen typically doesn't need any light to fall onto it for it to be visible to the observer. In fact, too much light falling on the VDT screen can make the screen difficult and/or uncomfortable to read.

For conversational VDT tasks, where the observer is "conversing" with the VDT screen typically there is no paper document. Therefore, there is little need for great quantities of light. Recognize, however, that spaces housing conversational VDT tasks could be virtually dark. Would such a space constitute a comfortable, preferable work environment? For most people, probably not—hence, the very real need to consider the subjective aspects of lighting, which will be addressed in Chapter 3.

Typically, guidelines written for electronic workplace lighting include some reference to illuminance. These references are usually made to recommended ranges of illuminance, since the tasks within the office can require from no light to enough light to read finely detailed pencil drawings or notes. Some guidelines propose ranges that include these extremes. Other guidelines propose ranges that include average tasks normally encountered in many, but not all, electronic offices. Caution is advised with any of these guidelines. The designer must be able to do one or more of the following prior to establishing a design target:

> To establish design target criteria, the designer needs to identify:
>
> ▼ Visual tasks to be performed
> ▼ The likelihood of workers encountering task extremes,
> ▼ The most frequent and/or most important visual tasks, and
> ▼ The intent of the design guideline being referenced

▼ **Identify the characteristics of the visual tasks expected to be performed in the environmental setting under consideration**—that is, make an assessment of the kinds of visual tasks that will be performed in the new or retrofitted setting. It is important to determine how detailed and difficult the visual tasks are to assess illuminance requirements. Typically, with the exception of reading VDT screens, more-detailed and difficult visual tasks require higher illuminances than less-detailed and easier visual tasks.

▼ **Identify the likelihood of workers encountering task extremes**—that is, those tasks that may require lots of light (fine detail drawings or notes; or poor xerography, for example) or those tasks that require little or no light (exclusively reading the VDT monitor as the sight task). If the likelihood of encountering such task extremes is high, then consideration should be given to making provisions to meet the illuminance targets for these extremes.

▼ **Identify what constitutes the majority and/or the most important visual work**—that is, of the tasks that have been identified, those that are performed most frequently and/or are most important. Illuminance criteria should be established that addresses these tasks.

▼ **Identify the intent of the guideline**—that is, is it attempting to address task extremes or task norms? Guidelines that are addressing task extremes should only be used when encountering task extremes, or the recommendations made by such guidelines should be tempered.

TABLE 2-2   *Visual Task Survey*

Visual Tasks	Anticipated Frequency			Anticipated Importance			Comments
	Lots	Fair Amount	Not Much	Great	Moderate	Little	
☐ Reading handwriting in ink	☐	☐	☐	☐	☐	☐	
☐ Reading handwriting in pencil	☐	☐	☐	☐	☐	☐	
☐ Reading printed matter							
☐ Small	☐	☐	☐	☐	☐	☐	
☐ Large	☐	☐	☐	☐	☐	☐	
☐ High contrast (crisp, black)	☐	☐	☐	☐	☐	☐	
☐ Low contrast (fuzzy, gray)	☐	☐	☐	☐	☐	☐	
☐ Writing in ink	☐	☐	☐	☐	☐	☐	
☐ Writing in pencil	☐	☐	☐	☐	☐	☐	
☐ Facial recognition	☐	☐	☐	☐	☐	☐	
☐ Accounting	☐	☐	☐	☐	☐	☐	
☐ Ledgers	☐	☐	☐	☐	☐	☐	
☐ Currency	☐	☐	☐	☐	☐	☐	
☐ Video display terminal/CRT[a]							
☐ Monochrome	☐	☐	☐	☐	☐	☐	
☐ Color	☐	☐	☐	☐	☐	☐	
☐ Video display terminal/LCD[b]	☐	☐	☐	☐	☐	☐	
☐ Other/describe	☐	☐	☐	☐	☐	☐	

[a]Cathode ray tube technology provides an internally lighted screen, much like a television.
[b]Liquid crystal diode technology requires external lighting in order to provide a visible screen.

Table 2-2 can be used during a walk-through of the client's existing facilities to better identify the characteristics of visual tasks, the likelihood of workers encountering task extremes, and what constitutes the majority and/or most important visual work activities. This review of existing conditions should also include measuring existing illuminances to know light levels to which the workers are accustomed. These measurements should not be used as design targets but as benchmarks against which to judge the validity of design decisions. Table 2-2 is reproduced in the back of the book for easy reproduction and use in surveying existing conditions.

If illuminance criteria targets are significantly different from existing illuminances, consideration should be given to reassessing illuminance target determinations (have the right judgments been made regarding the types of tasks, their frequency of occurrence and/or their importance?) and/or assessing the need for educating workers about the new lighting. This education process helps avoid the "shock" of very different lighting conditions without any explanation as to why such a drastic change has occurred.

Luminance

Luminance is the measured brightness of a surface, object, or area. This is the amount of light reflected from something or transmitted through something. Hence,

FIGURE 2-1

The negative-contrast VDT screen shown here "sees" a room setting (see Figure 2-3) that meets the luminance and luminance ratio guidelines discussed in Chapter 3. This yields a comfortable, acceptable electronic workplace. Figure 2-2 shows a positive-contrast VDT screen under the same lighting conditions. Courtesy of Fred Golden.

luminance is a "reaction" of a surface's reflectance and/or transmittance properties with a quantity of light (illuminance) falling on that surface. These characteristics are typically under the control of the designer.

By example, it is obvious why luminance of both task surfaces and room surfaces is so crucial a criterion to electronic workplace lighting. A VDT screen in negative-contrast mode is shown in Figure 2-1. Since the screen is negative-contrast and since the software is displaying in negative-contrast, the general luminance of the screen is relatively high. Near the top of the VDT screen several soft (not intense) veiling reflections are present that are caused by room surface luminances. Figure 2-2 shows the same VDT screen but with the software displaying in positive-contrast. Figure 2-3 shows the room view as seen by the VDT screen. Note that the library wall which runs from the suspended ceiling line (about 3.4-meters AFF) to the floor is lighted continuously top-to-bottom. This yields luminances on the library wall ranging from 9 to 50 cd/m² (the lower value of 9 registered at darker book binders and the higher value of 50 registered at lighter book binders). The luminance of the white bulkhead just above the shelves and continuing to the suspended ceiling line averages a uniform 50 cd/m². Above the suspended ceiling line is a cove uplight, which produces about 130 cd/m² near the suspended ceiling line. Finally, the luminaires produce about 950 cd/m² between 50° and 55° above nadir—the reference given to straight down (or 0°) below each luminaire (see Figure 3-1). In Figure 2-1, because the background of the text image on the VDT screen is light in color, and therefore high in luminance, the luminances of the room surfaces and luminaires are not interfering with viewing conditions.

FIGURE 2-2

A positive-contrast VDT screen is very sensitive to the room luminance conditions. In this case, veiling images usually cause accommodation problems, while veiling reflections cause adaptation problems. Either the screen needs to be changed (through software in this case) to a negative-contrast version like that shown in Figure 2-1, or the room surface and luminaire luminances need to be very low. Such low luminances, however, typically lead to a "cave-effect" work setting. Courtesy of Fred Golden.

On the other hand, Figure 2-2 illustrates the effects of the same room surface and luminaire luminances on a positive-contrast VDT screen. Note how the room surfaces show up slightly as veiling reflections. More disturbing, however, are the severe veiling reflections and veiling images caused by the luminaires. Because the screen background is dark, exhibiting no luminance, room surface luminances are more likely to be visible as veiling reflections and/or veiling images. In all of the figures in this example (2-1, 2-2, and 2-3), the illuminance on the work surface remained constant at 310 lux. Given the reflectances of the work surface and surrounding partitions, this illuminance level works with reflectances to yield acceptable luminances and luminance ratios in conjunction with paper tasks and negative-contrast VDT screens, as illustrated in Figure 2-1. This minimizes transient adaptation effects.

The reflections from the VDT screen in Figure 2-2 can cause at least two problems—an accommodation problem for the observer and/or a veiling reflection problem.

Accommodation is problematic because the eyes want to focus on the text or graphics on the VDT screen as well as the veiling images on the VDT screen. The text/graphics images have a focal length of perhaps 525 mm (21 inches) or so. The veiling images are reflections of objects which are probably more than 2 meters away. This multiple focusing can lead to headaches and tired eyes.

Veiling reflections are problematic because the reflections actually obscure the text-graphics on the screen. This can cause straining of the eyes or perhaps of the

FIGURE 2-3
This shows the room view as "seen" by the computer screen shown in Figures 2-1 and 2-2. Luminaires' and surfaces' luminances meet VDT-lighting criteria. Obviously the criteria should be considered maximum targets, with lower values being better— particularly evident in Figure 2-2. Courtesy of Fred Golden.

neck or back muscles as the head is tilted or the body reoriented to "see around" the veiling reflections.

The preceding discussion centers around luminance problems generated by electric lighting. Daylighting is as problematic, if not more so. Everyone loves to view out a window or see the sky overhead through a skylight. The romance of architecture leads many designers to mimic the large openings of medieval architecture. Modern interpretations include window walls, atria, and clerestories. Typically these daylight openings are problematic for electronic office workers because the openings have either inappropriate shading material or no shading material and/or they have inappropriate glass.

Shading material like blinds, shades, or drapes should be image preserving. That is, any material placed on or between window or skylight panes to help shade the brightness of the outdoor scenery should preserve the view of the window or skylight. Otherwise, why have a window or skylight? Exterior views are not only desirable because they connect people with the outdoor world but also because they assist in eye-muscle relaxation by providing distant views on which electronic office workers may focus periodically. Traditional horizontal or vertical blinds are typically opaque. No matter the position in which these opaque blinds are set, if any view to the exterior is preserved, then glare occurs because of excessive luminance. Sheer drapery material is also problematic, since it glows and causes a large white out condition. Image-preserving fabrics are available that provide as little as 3 percent light transmittance while still providing an exterior view. Figure 2-4 shows the visual effect of window walls with no window treatment, and Figure 2-5 shows the visual effect of window walls treated with 3 percent transmittance, image-preserving window covering.

Window and skylight glass typically exhibit very high luminance during daylight hours—in other words, glaringly bright. To put typical daylight luminances in perspective, Table 2-3 outlines some common daylight luminances and the impact of various window transmittances. An ordinary piece of white paper with text or some graphics on it may have a luminance of between 15 cd/m^2 and 240 cd/m^2, depending on task reflectance qualities and incident light intensity. The luminance of a computer screen may range between 6 cd/m^2 (text composed of light-colored or white letters on dark background) and 80 cd/m^2 (text composed of black letters on light-colored or white background).

Even with 30 percent transmission glass, an individual viewing the north sky or a VDT screen facing the window "sees" a luminance of 1,950 cd/m^2. This is twenty-four times brighter than some of the brightest computer screens. Such a luminance difference between screen and exterior view will cause distractions, glare, veiling reflections, and adaptation problems that could lead to visual fatigue and headaches, and most certainly will lead to reduced concentration and productivity.

Excessively bright surfaces, areas, and/or objects, including windows and sky-lights, can easily be avoided. Luminance criteria have been addressed by several worldwide guidelines. The issue of luminance, however, does not fit with the educational and practice background of many designers. As such, many designers and laypeople are intimidated by luminance criteria and the way to addressing them. Window glass transmittances of 5 percent to 10 percent surely must render interiors dark, dingy, and gloomy while presenting the exterior scene as a heavily overcast day, right? Skylight glass transmittances of 3 percent to 10 percent must create similarly gloomy interiors, right? Not necessarily. The design process needs to be more rigorous in exploring glazing and treatment alternatives that will result in satisfactory environments; rather than simply choosing the darkest-tint glazing a manufacturer can supply, the designer needs to evaluate the opportunities offered by image-preserving frit coatings and window shade treatments in conjunction with spectrally neutral glazings.

Lighting equipment has advanced to such a state that bright, glary luminaires are no longer a necessary evil. Glazing tints and coatings have advanced to such a state that very low transmission glass need not be dark gray and need not radically alter the color-rendering properties of incoming daylight (technically referred to as spectral transmission).

FIGURE 2-4

Even though this east-facing window wall has a transmittance less than 15 percent, without any interior window treatment, the luminance of the exterior sky and nearby buildings and landscaping are too bright—causing glare and silhouetting. Courtesy of Robert Eovaldi.

FIGURE 2-5

With an image-preserving shading material having a transmittance of just 3 percent (for a total window-system transmittance of just 1.5 percent), glare and silhouetting have been eliminated while view is preserved. Courtesy of Robert Eovaldi.

TABLE 2-3   *Empirical Sky Luminances (Based on Measurements in Ann Arbor, MI on December 28, 1993[a])*

Viewing Orientation	Outdoor Luminance	Interior Luminance with 70% Window or Skylight Transmittance	Interior Luminance with 30% Window or Skylight Transmittance	Interior Luminance with 5% Window or Skylight Transmittance
☐ North sky				
☐ Near horizon	☐ 6,500 cd/m²	☐ 4,550 cd/m²	☐ 1,950 cd/m²	☐ 325 cd/m²
☐ Directly overhead (deep blue)	☐ 980 cd/m²	☐ 686 cd/m²	☐ 294 cd/m²	☐ 49 cd/m²
☐ South sky near horizon	☐ 18,000 cd/m²	☐ 12,600 cd/m²	☐ 5,400 cd/m²	☐ 900 cd/m²
☐ Ground snow cover	☐ ranged 6,000 to 11,000 cd/m²	☐ 4,200 to 7,700 cd/m²	☐ 1,800 to 3,300 cd/m²	☐ 300 to 550 cd/m²
☐ Ground cover	☐ 1,600 cd/m²	☐ 1,120 cd/m²	☐ 480 cd/m²	☐ 80 cd/m²
☐ Medium-value brick pavers	☐ 2,500 cd/m²	☐ 1,750 cd/m²	☐ 750 cd/m²	☐ 125 cd/m²
☐ Stand of deciduous trees[b]	☐ 650 cd/m²	☐ 455 cd/m²	☐ 195 cd/m²	☐ 32 cd/m²

[a] *2:30 PM EST, mostly clear, some light haze on southern horizon. Measurements made with Minolta 1° Luminance Meter.*
[b] *Stand of trees face south, measurement taken viewing north so that trees are in full sunlight.*

Luminance Ratios

Absolute luminance or measured brightness, as discussed earlier, can be problematic but can be limited with appropriate design criteria and techniques. Even when luminances are brought within criteria ranges, they can still present a problem for electronic office workers. When one surface or area or object is significantly brighter or darker than an adjacent surface, area, or object, the luminance difference or ratio between the two can cause annoying and/or veiling reflections from VDT screens. A classic example is that of a white blouse or shirt reflecting from the VDT screen. The luminance of the blouse or shirt on a worker in a work setting with carefully designed lighting likely does not exceed any criteria limits. But if the furniture background finishes are dark, the white blouse or shirt stands out and reflects from the VDT screen.

Another example of the importance of luminance ratio criteria occurs with indirect lighting systems. Although the brightest spots on a ceiling may not exceed luminance criteria, if these spots are more than six or seven times brighter than the adjacent darkest spot, then this luminance difference or ratio will reflect from the VDT screen causing veiling reflections or "washout." Frankly, in a perfect viewing environment for a VDT screen, everything in the room would have the same luminance. This is why so many people simply opt to work in a "cave." By turning off all electric lighting, except a task light perhaps, nearly everything in the room has zero luminance and more important no luminance differences or ratios to cause computer screen washout. Of course, the environment is like a black hole, so that psychologically it has a closed-in, somewhat oppressive feel. More important, however, for those folks working on positive-contrast screens, the luminance ratio between the screen and most of the surroundings is now out of whack. This can contribute to visual fatigue, headaches, and, ultimately, loss in productivity. Finally, if a task light is used for reading hard-copy materials, the luminance ratio between the lighted task and the darker background is too great.

Luminance and luminance ratio criteria are the heart of the problem. These criteria must be aggressively targeted by designers, engineers, owners, and workers if the electronic office is to be a success—a comfortable, productive work setting.

Surface Reflectances

Luminances and luminance ratios experienced in the work environment are in part a result of the reflectances of the environmental surfaces and objects. Typically, reflectance criteria are reported in ranges, since it is undesirable to have surfaces too low in reflectance (contributing to a cave effect) and/or to have surfaces too high in reflectance (contributing to luminance ratio problems if immediately adjacent to surfaces of lower reflectances).

Very low surface reflectances, typically less than 10 percent in high-light-level settings and less than 20 percent in low-light-level settings, contribute to transient adaptation problems and glare problems. Scanning a view around a room with such low reflectances while working on paper documents under rather high-light-level settings will lead to delayed work resumption as the eyes adapt from one brightness extreme to another and back again. The dark background contrasts harshly to lighted paper tasks and/or to negative-contrast VDT screens, resulting in glare sensation—perhaps of such low grade that the only manifestation may be a headache later in the day. Hence, low reflectance surfaces should be minimized.

Very high surface reflectances, typically exceeding 50 percent on the walls and 75 percent on the ceiling in moderate- to high-light-level settings, also contribute to transient adaptation problems and glare problems. Scanning a view around a room with such high reflectances while working on a positive-contrast VDT screen will lead to delayed work resumption as the eyes adapt from one brightness extreme to another. The bright background contrasts harshly to positive-contrast VDT screens, resulting in glare sensation.

A midrange of reflectances for surfaces helps provide the potential for a work setting that offers comfortable working conditions for those using VDT screens. All surface finishes should be matte (dull or diffuse) rather than specular (shiny or mirrorlike). Manufacturers of almost all interior building surfaces (paints, laminates, ceilings, carpets, etc.) have reflectance values for their respective surfaces. Manufacturers of furniture systems typically have reflectance values for their respective furniture elements. If all else fails, some simple measurements can be made with an inexpensive illuminance meter in just a few minutes of time to roughly establish surface reflectances; the material in question, however, needs to be available in samples as large as a letter-size sheet of paper. As illustrated in Figures 2-6 and 2-7, use an ordinary sheet of white copier paper or computer-printer paper as a "reference"

FIGURE 2-6

A reference sheet of white paper and a piece of the material for which a reflectance value is to be determined are placed side by side in an open area of uniform, moderate-to-high-level lighting. The light meter is aimed so that the white photocell is directed toward the white reference paper and held about 100 mm (4 inches) above the paper and a reading is taken. Repeat the procedure for a piece of the material (see Figure 2-7). Courtesy of Fred Golden.

FIGURE 2-7

Repeat the procedure outlined in Figure 2-6 for the sample of material in question. Use Equation 1 to estimate the reflectance of the material. To maintain SI consistency, all values of illuminance read from the meter were converted to SI using 10.76 lux to 1 foot-candle (multiplying the value read from the meter by 10.76). Courtesy of Fred Golden.

reflectance sheet. In an area of relatively high, uniform lighting (could be done out-doors in daylight), place the white reference sheet and the sample of the material in question side by side faceup on a flat, horizontal surface (floor, table, etc.). Then place the illuminance meter about 100 mm (4 inches) above the white reference paper with the illuminance meter photocell aimed toward the paper. Take the illuminance reading (so many lux or footcandles). Now repeat the procedure for the sample material without moving any of the materials and staying in the same physical position so that the lighting conditions remain identical for both readings. Take the illuminance reading of the light reflected from the sample material. Mathematically, the surface reflectance of the material in question can be estimated as follows:

EQUATION I

$$Material\ Reflectance\ (\%) = \frac{70\% \times Illuminance_{material}}{Illuminance_{paper}}$$

Where $Illuminance_{material}$ = amount of illuminance measured as reflecting from the material in question; and where $Illuminance_{paper}$ = amount of illuminance measured as reflecting from the white paper. If the light reflected from the paper ($Illuminance_{paper}$) is measured at 380 lux; and the light reflected from the material sample ($Illuminance_{material}$) is measured at 160 lux, then:

$$Material\ Reflectance\ (\%) = \frac{70\% \times 160\ lux}{380\ lux} = 30\%$$

Power Budget

Resolving lighting issues for people using computers must simultaneously minimize the impact on earth's environment. Lamp, ballast, and luminaire technologies are such today that any good design (that is, a design that meets the viewing needs of people using computers) should not have an excessive connected load. In fact, high lighting loads (high wattage loads) will likely lead to user discomfort because of increased cooling loads. In other words, lighting solutions which use a lot of energy will make the environment hot—a situation compounded by the high wattage computers with which people are working. No lighting solution is complete without addressing the lighting system power budget.

Subjective Impressions

Psychological or subjective issues of lighting for people using computers are not well understood or fully researched. Nevertheless, as discussed previously in this text, such subjective issues as the cave effect are experienced by many people using computers in essentially dark rooms, where the only light is that of a task light oriented at reference paper documents. Lighting research in the mid-1970s fueled by prospects of low light levels resulting from energy conservation measures indicated that luminance intensity, uniformity, and distribution appear to affect people's subjective reactions to visual environments.[22] Subjective impressions that seem to be somewhat influenced by the luminance variables cited above include visual clarity, spaciousness, relaxation, and intimacy. Unfortunately, none of the cited guidelines address psychological issues.

Miscellaneous Issues

Various other issues cannot be conveniently categorized with those issues previously identified. Additionally, although the previously identified issues are addressed nearly unanimously by design drafts, guidelines, and standards around the world, other issues exist that are deemed important by some but not by others. Thus the role of the designer becomes most significant in identifying those issues that are important for a given client's set of circumstances.

These miscellaneous issues include: luminaire luminances; application issues of task lighting, direct lighting, direct/indirect lighting, and indirect lighting equipment; color temperature; and employee education programs regarding the use of VDTs.

Other issues of importance include the following:

▼ Luminaire luminance,

▼ Application issues related to task lighting, direct lighting, direct/indirect lighting, and indirect lighting,

▼ Color temperature, and

▼ Employee education programs

It is unfortunate that luminaire luminances are not universally recognized as a significant issue in the electronic workplace. As important as, if not more important than, surface luminances and luminance ratios, luminaire luminances can singlehandedly wreak havoc in the electronic workplace. Direct or direct/indirect luminaires that are too bright (luminaire luminance is too high) can obliterate text on a VDT screen and can cause: veiling reflections, annoying reflected imaging from VDT screens, direct glare, and reflected glare.

Application issues are addressed in some guidelines with prescriptive measures. Illuminances, luminances, luminance ratios, and surface reflectances are primarily performance issues. If the designer develops an interior work environment that meets these performance criteria, then generally the environment should provide the potential for good to excellent visibility of VDT screens by workers. Prescriptive measures or criteria, however, offer a prescribed lighting solution; or, by banning certain prescribed solutions, limit the designer to certain choices. Yet prescribing solutions is not likely to lead to improved results. For example, the New Jersey State *Guidelines for the Use and Functioning of Video Display Terminals, Part I*, recommends that indirect lighting be used when/where possible. Such a prescription, without attaching performance criteria of luminances and luminance ratios, can just as easily lead to inappropriately overlighted and/or glary electronic workplaces as might the use of direct lighting. Decreeing solutions without pushing designers to figure out the question will surely lead to bad designs.

Employee education programs, like luminaire luminance criteria, should be a universally recognized issue of any electronic workplace design. Many times employees exacerbate the problems with viewing of VDTs because the intuitive thing to do is at complete odds with the correct thing to do. It is intuitively appropriate to work on a computer in a darkened room. Therefore people want to turn out electric lights and close blinds or drapes. Yet these very actions can lead to just as much visual fatigue as the "bad" electrically lighted, daylighted environment.

Luminaire luminances, if not within limits, can:

▼ Obliterate text on a VDT screen.

▼ Cause veiling reflections,

▼ Cause annoying veiling images,

▼ Cause direct glare, and

▼ Cause reflected glare

Intuitively, to "keep plugging away at the keyboard" will result in more work completed, so many folks believe. On the contrary, visual breaks, mental breaks, and breaks that are taken in areas or spaces physically separated from the work area can help "rejuvenate" the eye and hand muscles and relax the neck and back muscles. Employees must be made aware of these and other aspects of working on VDTs. In-house training programs, video programs, or a monthly or quarterly newsletter program can help educate and periodically remind people of the best methods for maintaining comfort and ultimately productivity.

Subjective issues that are generally difficult or impossible to quantify are only briefly addressed by a few drafts, guidelines, or standards. Of particular note is the lack of discussion regarding workers' subjective impressions and/or reactions to the work setting. The thought of coming to work or returning from break only to find a small, dark hole in which to work for the next several hours may not appropriately motivate many people. The issue of pleasantness is not addressed by any of the cited documents.

Finally, lighting, although a significant part of how people perceive, react, and produce in a workplace, is only one of many other environmental, architectural, and interior systems that need to be addressed. Acoustics and/or thermal comfort in a workplace can limit or enhance productivity regardless of the lighting. Physical layout and support of work functions can limit or enhance productivity. These issues and others are addressed by many of the referenced documents and/or their respective authoring bodies listed in Table 2-4.

Surveying Worldwide Developments

Lighting is as much an art as a science. Not enough is known, and likely never will be known, about lighting to justify an all-quantifiable approach to lighting design. There are too many variables, most notably, every person is different from every other person. Attempting to assess the significance of lighting in such a way as to generate criteria with decimal-point accuracy, which in turn will generate environments of guaranteed satisfaction and productivity, is beyond practical research methods, and financing, not to mention that it would be a never-ending task. Because as tasks, people, and societal values change, so too change people's attitudes, behavior, productivity, and lighting requirements. As such, lighting guidelines tend to vary, in some cases significantly, around the world and through time.

Many times, employees exacerbate the problems associated with viewing of VDTs because the intuitive thing to do is at complete odds with the correct thing to do.

In cursory fashion, Table 2-4 outlines a survey of some of the worldwide design drafts, guidelines, and standards regarding lighting for people in electronic workplaces that have been available within the past decade. Snippets of information are provided to simply introduce the reader to the variety and magnitude of lighting criteria. For specific, definitive guidance from any of the drafts, guidelines, or standards cited, reference should be made to those documents in their entirety (complete reference is given in the Endnotes). Because of changing technologies and ideologies, some of the surveyed material is already dated and is being deprecated or updated. At least one of the cited documents, *Health Code of the San Francisco Municipal Code*, Part II, Chapter 5, Article 23, has been declared unconstitutional in its 1990 form. It is reasonable to presume that all of these references could be dated within a few years. The reader is therefore encouraged to obtain the latest versions of the referenced material.

TABLE 2-4A   *Survey of Selected Worldwide Lighting Criteria*
Boeing[23]

Criteria	Design Targets
Illuminances	▼ Ambient of 300 lux; with preference for adjustment from 50 to 500 lux ▼ Ambient of 500 lux where extensive reading and writing of paper materials occurs; with preference for adjustment from 100 to 1,000 lux ▼ 2,500 lux on paper reference materials; with preference for adjustment from 500 to 5,000 lux
Surface Luminances	▼ Limit maximum to twice the VDT image luminance [not practical with positive contrast screens, particularly with respect to lighting paper tasks to more than 300 lux] ▼ Limit minimum to one-tenth VDT image luminance
Luminance Ratios	▼ See Surface Luminances
Surface Reflectances	▼ Ceiling: 80–92% ▼ Walls: 40–60% ▼ Floors: 21–39% ▼ Furniture: 26–44%
Power Budget	Not addressed
Subjective Impressions	Not addressed
Miscellaneous	None relevant to this text discussion

TABLE 2-4B   *Survey of Selected Worldwide Lighting Criteria*

International Commission on Illumination (Commission Internationale de l'Eclairage or CIE)[20]

Criteria	Design Targets
Illuminances	▼ 300 to 1,000 lux on horizontal work plane
Surface Luminances	Not addressed
Luminance Ratios	Not addressed
Surface Reflectances	▼ Ceiling: ≥70% (matte)
	▼ Walls: 50–70%
	▼ Floors: 30%
	▼ Window treatment: 50–70%
Power Budget	Not addressed
Subjective Impressions	Not addressed
Miscellaneous	▼ For direct lighting, luminaires should have cutoff angles in the range of 45°–55°, with maximum luminance at cutoff of 200 cd/m^2
	▼ For indirect lighting, look for widespread distribution and uniform, matte ceiling reflectance

TABLE 2-4C   *Survey of Selected Worldwide Lighting Criteria*

The Chartered Institution of Building Services Engineers (CIBSE)[19]

Criteria	Design Targets
Illuminances	▼ 300 lux on digitizer
Surface Luminances	▼ ≤500 cd/m^2 for surfaces directly in front of VDT screen
	▼ ≤1,500 cd/m^2 for walls and ceilings with only "gradual" variation
	▼ ≤500 cd/m^2 average for ceiling for indirect lighting with maximum of 1,500 cd/m^2
	▼ ≤70 cd/m^2 on board of graphic workstation with all other surface luminances ≤200 cd/m^2
Luminance Ratios	Not addressed
Surface Reflectances	▼ Ceiling: ≥70%
	▼ Floor: high
Power Budget	Not addressed
Subjective Impressions	Not addressed
Miscellaneous	▼ For direct lighting in VDT-intensive areas, luminaire luminance should be <200 cd/m^2 above a cutoff angle of 55°
	▼ Indirect lighting eliminates harsh reflections
	▼ Mounting height of task lights should be <½ the width of task area to be lighted

TABLE 2-4D   *Survey of Selected Worldwide Lighting Criteria*
German Institute for Norms (Deutsches Institut für Normung or DIN)[16]

Criteria	Design Targets
Illuminances	▼ Ambient of 500 lux
Surface Luminances	▼ <200 cd/m² mean luminance of all surfaces with a maximum ≤400 cd/m²
Luminance Ratios	Not addressed
Surface Reflectances	Not addressed
Power Budget	Not addressed
Subjective Impressions	Not addressed
Miscellaneous	▼ For direct lighting, luminaire luminance should be <200 cd/m² in the cutoff zone ▼ Task lights are discouraged unless direct and reflected glare are avoided and luminance ratios not increased [this seems impossible, given that a task light will put more light onto a paper task with hopefully no light on the VDT task—hence ratios will always be increased]

TABLE 2-4E   *Survey of Selected Worldwide Lighting Criteria*
Illuminating Engineering Society of North America[18]

Criteria	Design Targets
Illuminances	▼ Ambient of 300 to 500 lux providing VDT screen luminance is ≥50 cd/m² (that is, the VDT screen, when energized with text has a luminance of 50 cd/m² or greater, excluding any electric light or daylight reflections)
Surface Luminances	▼ ≤850 cd/m² average from the ceiling over any zone of 0.35 m²
Luminance Ratios	▼ Paper to VDT = 3:1 ▼ VDT to nearby dark surface = 3:1 ▼ VDT to nearby light surface = 1:3 ▼ VDT to remote dark surface = 10:1 ▼ VDT to remote light surface = 1:10 ▼ Ceiling directly over luminaire to ceiling between luminaires (for indirect lighting) ≤4:1
Surface Reflectances	Not addressed directly (addressed through Surface Luminance and Luminance Ratio Criteria)
Power Budget	Not addressed
Subjective Impressions	Not addressed
Miscellaneous	▼ For direct or direct/indirect lighting, luminaires should have maximum luminances of 850 cd/m², 350 cd/m², and 175 cd/m² at cutoff angles of 55°, 65°, and 75° respectively.

TABLE 2-4F   *Survey of Selected Worldwide Lighting Criteria*
International Organization for Standardization (ISO)[24]

Criteria	Design Targets
Illuminances Surface Luminances	Recommendations based on luminances ▼ Average luminance ≤ 200 cd/m^2 for surfaces surrounding the VDT screen ▼ Maximum luminance of 400 cd/m^2
Luminance Ratios	Not addressed
Surface Reflectances	▼ Ceiling: 60–80% ▼ Walls: 40–80% ▼ Floors: 15–25% ▼ Furniture: 20–50% ▼ Paper document: 40–80%
Power Budget	Not addressed
Subjective Impressions	Not addressed
Miscellaneous	▼ For direct lighting, luminaires should have luminances of <500 cd/m^2 at angles above 45°, with preferable luminances of 200 cd/m^2 to 400 cd/m^2 ▼ For indirect lighting, average ceiling luminances should not exceed 500 cd/m^2 ▼ Consider indirect or direct/indirect lighting ▼ Achieve uniform brightness distribution ▼ Consider individual, controllable task lighting in conjunction with general lighting (do not use local task lighting exclusively)

TABLE 2-4G   *Survey of Selected Worldwide Lighting Criteria*
State of New Jersey[17]

Criteria	Design Targets
Illuminances	▼ Ambient of 200 to 700 lux
Surface Luminances	Not addressed
Luminance Ratios	Not addressed
Surface Reflectances	Not addressed
Power Budget	Not addressed
Subjective Impressions	Not addressed
Miscellaneous	▼ Indirect lighting is recommended ▼ If direct lighting is used, then parabolic louvers are recommended ▼ In existing systems, consider relamping with lower output lamps and retrofitting lens/louvers

TABLE 2-4H   *Survey of Selected Worldwide Lighting Criteria*
Public Works Canada[25]

Criteria	Design Targets
Illuminances	▼ *Ambient of 200 lux*
	▼ *Task of 300 to 750 lux*
Surface Luminances	▼ *Not addressed*
Luminance Ratios	▼ *Paper to VDT ≤ 3:1*
	▼ *VDT to nearby dark surface ≤ 3:1*
	▼ *VDT to nearby light surface ≤ 1:3*
	▼ *VDT to remote dark surface ≤ 10:1*
	▼ *VDT to remote light surface ≤ 1:10*
	▼ *VDT to window ≤ 1:20*
	▼ *VDT to ceiling ≤ 1:20*
	▼ *Ceiling average to ceiling minimum ≤ 3:1*
Surface Reflectances	Not addressed
Power Budget	Not addressed
Subjective Impressions	Not addressed
Miscellaneous	▼ *Direct lighting with parabolic louvers results in dark walls and ceilings; may wish to correct with wall lighting or some indirect lighting*
	▼ *3000°K sources perceived as less glary [this is anecdotal experience]*

TABLE 2-4I   *Survey of Selected Worldwide Lighting Criteria*
City of San Francisco[26]

Criteria	Design Targets
Illuminances	▼ *Ambient of 200 to 500 lux*
Surface Luminances	Not addressed
Luminance Ratios	Not addressed
Surface Reflectances	Not addressed
Power Budget	Not addressed
Subjective Impressions	Not addressed
Miscellaneous	▼ *Task light(s) to be provided at the request of occupant*
	▼ *Employers must inform and train VDT users regarding the vision effects of VDT work and protective measures including: regular breaks from VDT; mechanisms for reducing glare; vision exams; eye exercises*

3 | A Model Lighting Guideline

Some significant variations exist among several of the guidelines surveyed in Table 2-4. Clearly, lighting for people using computers greatly depends on the kind of VDT task, the kind of paper task, and the frequency of performance of each of those tasks. The designer must know the kinds of tasks for which lighting solutions are to be developed. Simply put, there are no easy prescriptive rules of thumb. The designer must form the correct "question"—that is, understand the problem—before reasonable solutions can be developed and implemented.

Nevertheless, once beyond establishing the visual tasks, the designer has a myriad of criteria resources. As Table 2-4 indicates, some of these resources only address a few issues. Some seem to be in conflict with others. Some espouse very similar criteria.

Why a Model Guideline?

The development of any draft, guideline, or standard for any issue is quite complex. In democratic societies, groups of experts volunteer or are paid to develop a consensus approach. Although this ultimately leads to study of a wide range of issues, it typically requires a lengthy process of compromises. Depending on the intent of the group, these compromises can lead to documents that provide some guidance for some designers for some applications some of the time. This is predominant for groups with very large and diverse constituencies. On the other hand, groups composed of members from small and similar constituencies can develop guidance for a few designers for a few applications all of the time. Still others may have compromised on the depth as well as breadth. Though all of these groups are well intentioned, by nature of each group's charge and composition, the outcome is likely to be compromised.

A model guideline can build on experience. Those elements of diversity from various reference guidelines can be "tested." Those elements of similarity can be employed on a regular basis and "stretched" or "tightened" depending on experience. A model guideline can help designers find a common base from which to develop lighting solutions to meet the problems identified by the guideline as important and acknowledged by the designer as existing on a given project. A model guideline does not and cannot usurp local, state, or national codes or ordinances.

What Are the Tasks?

Understanding the kinds of work people do in an electronic workplace is sometimes straightforward and other times quite complex. Unfortunately, this cannot be ascertained without reviewing the existing tasks that people do in their workplace. The visual task survey of Table 2-2 not only helps the designer establish the visual tasks performed but also can help establish their hierarchical importance and how much of the time they are performed. If reading and writing in pencil isn't performed much and isn't terribly important, then it doesn't make sense to design a lighting system that provides enough light of the right quality for reading/writing in pencil. Once this visual task survey is completed, it should be reviewed by the client. Since the client may know of pending new kinds of work tasks that may require new and different kinds of lighting, this review is important if not critical.

Most people work on computers at some point during the day. The designer must establish what kind of VDT is used, how often is it used, and what is the significance of its use (how important it is to the function of the worker and to the employer). Additionally, the designer must establish what other types of visual tasks are performed. Once all of the visual tasks are understood, the designer can review criteria guidelines and establish those that are most relevant to the situation at hand.

CADD VDT Tasks

Computer-aided design and drafting (CADD) tasks are perhaps the most difficult for which to provide proper lighting. All CADD tasks typically involve one, several, or all of the following visual tasks:

▼ Visually inspecting and assessing fine line work on positive-contrast VDT screens;

▼ Visually inspecting and assessing fine line work on negative-contrast VDT screens;

▼ Visually inspecting and assessing fine line work on blueprints;

▼ Visually inspecting and assessing fine line work on original drawings (pen or pencil);

▼ Reading instructions from books, reports, memos, or handwritten notes or sketches

Any one of these visual tasks alone demands visual scrutiny unparalleled by any other officelike visual task. Since most CADD tasks combine the preceding visual tasks, the lighting must be quite carefully studied and developed if reasonable accuracy, comfort, and productivity are goals.

Typical CADD tasks involve reading both hard copy and VDT screen images. Comparing line work and dimensions of detailed images on VDT screens to hard copy references and vice versa is demanding because of the extent of the detail that must be reviewed and because of the extent of viewing between hard copy and VDT screen.

Conversational VDT Tasks

Conversing with a VDT only recently has come to mean literally speaking with the computer. This technology, however, remains a future possibility for most VDT users. Typically, "conversing" with a VDT means communicating with the computer via typewritten commands, with the computer responding either via text or graphic images on a VDT screen or text or graphic images' printed output or both. Many "conversational" VDT tasks do not require much, if any, reference to paper documents during the process of conversing. For example, telephone operators, airline reservationists, mail-order salespeople, and insurance agents spend much of their workday taking telephone inquiries from clients, translating these telephone inquiries into typewritten instructions to a computer, which in turn responds with information via the VDT screen. This information is then conveyed to the clients via telephone. The process is quite iterative—repeating itself until either the client's and/or the computer user's inquiries are satisfied.

Another example of conversational VDT tasks that are common to many white-collar support jobs is transcription of taped testimony, taped conversations, and dictation. Very little reference is made to any paper tasks. The worker listens to the audiotapes and transcribes the information to written form via the computer.

Since the conversational VDT tasks do not require much work with or reference to hard-copy documents, the primary visual task is viewing the VDT screen. As such, the lighting criteria can be more specifically intended to provide optimum viewing conditions of the VDT screen itself.

Data-Entry VDT Tasks

Some VDT tasks require the worker to input hard-copy information to the computer. This input material can be in the form of numerical information and alphanumeric information, commonly referred to as "data." This includes accounting ledgers, correspondence, reports, books, and the like.

Data-entry tasks usually require the worker to read hard-copy information on a continual basis, with intermittent scans of the VDT screen to assess the accuracy and status of the information being input to the computer. For data-entry VDT tasks, the lighting criteria are typically intended to provide optimum viewing conditions of the paperwork, sometimes allowing viewing conditions of the VDT screen to be less than optimal.

Combination Tasks

Some people's work involves several of the previously mentioned electronic tasks. Understanding the variation of these tasks and the amount of time spent on each task during a typical workday is important to designing the lighting.

Negative-Contrast VDT Monitors

Negative contrast refers to the presentation of relatively dark text and/or graphic images on light backgrounds. This book is a negative-contrast task (essentially black ink on white paper). The VDT monitors and the support software available today readily provide negative-contrast images. These are typically preferable to positive-contrast VDT monitors, since room lighting conditions and light reflections are not as problematic in washing out VDT screen images. Negative-contrast monitors, however, can have two serious faults—flicker and too-high inherent brightness. For flicker to be unnoticeable, the monitor needs to be a noninterlaced-type screen operating with a refresh rate of at least 72 Hz, and preferably at 90 Hz.

For older users, the negative-contrast VDT monitors can have too-high inherent brightness, that is, the light-colored background glows too brightly. This can introduce light scatter on the retina, thereby reducing task visibility and increasing discomfort. Lighting for people using computers is a difficult chore. Fine-tuning of the software to achieve the appropriate balance between negative-contrast screen and reference papers and the overall environmental setting will likely be necessary in many situations for older users.

Positive-Contrast VDT Monitors

The early VDT monitors typically presented light-colored or luminous information (text or graphic images) on a dark background. Although technically referred to as positive-contrast VDT monitors, these are known more popularly as monochromatic screens. The classic green or amber screens remain active in many workplaces today. Although the inherent task contrast is very good with these kinds of screens, they do not interact well with a lighted environment. For example, the monitor glass with the dark background tends to act just like a mirror. Hence, the more light that is introduced into a space, and the more lighted or bright objects become, the more likely are screen veiling reflections, screen glare, or screen washout.

Since positive-contrast VDT monitors are much more susceptible to lighting problems, the lighting for workplaces with these kinds of monitors must be more carefully designed. Strict adherence to lighting criteria is a necessity. Additionally, positive-contrast VDT screens do not lend themselves well to working on a combination of tasks. Viewing from hard-copy documents to darkened VDT monitors can cause adaptation effects to compound visual fatigue and headaches. Negative-contrast VDT monitors offer advantages that are typically preferred over the positive-contrast VDT monitors.

What Are the Key Criteria?

Criteria related to how well people can see in an electronic workplace can be categorized as lighting criteria and nonlighting criteria. Much of the nonlighting criteria has already been addressed in Chapter 1, including *Computer Screen, Screen and Monitor Maintenance,* and *Paper Document.* Some lighting criteria also have been discussed throughout Chapters 1 and 2. This section outlines all lighting criteria that should be considered in the design of lighting for electronic workplaces.

Lighting criteria depend on the visual tasks expected to be performed. In fact, perhaps one of the more significant application and design oversights that designers can make is to apply a single lighting criterion across the board to a host of visual tasks. As such, the lighting criteria that follow are based on the visual task(s) that is/are anticipated to be performed by the users.

It is incumbent on the designer and/or user to identify the anticipated visual tasks, the frequency with which they are performed, and the extent of their significance to the users' overall performance. The electronic office tasks have been previously categorized as CADD VDT, conversational VDT, data-entry VDT, and a combination of any or all of these. Ultimately, the designer needs to either merge criteria required for the successful and comfortable performance of the various tasks; select criteria that favor one task that is identified as "most critical" or performed "most frequently"; or develop several sets of criteria, and subsequently several sets of lighting conditions or "scenes," that can be independently operated to meet criteria for each task as it occurs.

Lighting criteria can be grouped by six basic topics: illuminances, luminances, surface reflectances, power budget, subjective impressions, and maintenance. Within each topic several subsets of criteria need be addressed. Illuminance issues include ambient horizontal illuminances, ambient vertical illuminances, and task illuminances. Luminance issues include absolute luminances of surfaces and objects, luminance ratios, and luminaire luminances. Maintenance issues include cleaning and relamping. Of these criteria, the most valuable are the luminance criteria. If illuminance and/or maintenance criteria are used to the exclusion of luminance criteria, it is quite likely that the lighting results will be seriously compromised and unsatisfactory for the people using computers.

Illuminance

Illuminance is a misused criterion in lighting design. Yet because of the historical significance of horizontal illuminance (e.g., the amount of light on a work surface), its ease of use in calculations and measurements, and its being the only criteria cited, if any are cited, in popular-press articles, it remains for many designers the sole lighting criterion. There is no question that illuminance is a significant criterion, but more than just horizontal illuminance must be evaluated. Most important, illuminance criteria must be used in conjunction with such criteria as luminance, uniformities, and the like to establish comfortable, productive viewing conditions.

Oddly, if a complete series of illuminance criteria were applied sensibly and rigorously on every project, then illuminance criteria might suffice for many projects. Apparently, however, ever-increasing time pressures and economic pressures push designers toward "single-number" criterion—and for years horizontal illuminance has stood out as the most convenient. Now is the time to reconsider the implications of using a single-value criterion for designing lighting that is to make spaces more user-friendly and the environment more earth-friendly.

Illuminance criteria only make sense if applied holistically; that is, with respect to the total environmental setting and its impact on comfort and productivity. This means reviewing not only illuminance requirements necessary for reading of tasks but also illuminance requirements to reduce visual fatigue caused by adaptation effects.

For both energy conservation and visual variety, it is seldom desirable to design high levels of general lighting throughout a space to accommodate any task anywhere. Also, as computer tasks have increased in popularity, it has become obvious that high levels of general or ambient light are undesirable. Illuminance requirements are therefore best implemented in layers—a layer of ambient or general horizontal illuminance, of ambient vertical illuminance, and of task illuminance.

Ambient Horizontal Illuminance

The amount of ambient or general light falling on horizontal surfaces, including the work plane, is typically and inappropriately the only lighting design criterion used by many designers. Decades ago this was acceptable practice. Most tasks were paper tasks, and most of these were of relatively poor contrast (hard pencil on paper, mimeograph, early xerograph, and so on). For such poor-contrast paper tasks, lots of light was needed. Additionally, electricity was cheap and environmental concerns were nonexistent. Prismatic-lensed luminaires were stuffed with four lamps and threw a lot of light around a room; enough hit the walls and bounced around off of the horizontal work surfaces that the rooms appeared rather bright if not downright glary. Task lights (table or wall lights that are used for the specific purpose of lighting the task) were unnecessary.

> Ambient horizontal illuminances that are too high can lead to adaptation problems and visual distractions caused by brightly lighted surroundings and objects . . . leading to visual fatigue, headaches, and reduced productivity.

Today many tasks are of inherent high contrast. Black felt-tip pen on white or light-yellow paper, inkjet- and laserjet-printed documentation, crisp xerography, and so on compose many paper tasks. These tasks are quite visible with 300 lux on them, and only a few paper tasks need more than 750 lux on them for most people to see them well.

With conversational VDT and some CADD VDT tasks, the worker may periodically view back and forth between a reference paper task and the computer screen. Additionally, just like any other demanding visual work, visual breaks are desirable—viewing around the space or out a window for several minutes every half hour or so. If the surrounding nearby horizontal surfaces are too bright or too dark and/or the computer screen is too dark or too bright, respectively, then this ratio or luminance difference is significant enough to cause adaptation problems. Horizontal ambient illuminance criteria guidelines can help minimize this problem, as can the use of negative-contrast computer screens. If the computer screen exhibits a light-colored background with dark text/graphics (negative contrast), then it is usually reasonable to light the surrounding horizontal surfaces up to an average of 375 lux. For CADD and conversational VDT tasks, however, such a value should be reserved for those older viewers who have a task requiring frequent referencing of the paper document. Table 3-1 outlines suggested ambient horizontal illuminances for a variety of CADD or conversational VDT tasks and users' conditions.

In any event, the horizontal surfaces (work surfaces and floor) should have reflectances as recommended later in "Surface Reflectances." If surface reflectances are too low or too high, then the ambient illuminance guidelines suggested here may be too low or too high, respectively.

With data-entry VDT tasks, the worker's view alternates between reference paper documents and the computer screen with much time spent viewing reference paper documents. At the same time, viewing around the space or out a window for several minutes every half hour or so is a desirable visual break activity. If the computer screen is negative contrast, then it is reasonable to light the surrounding horizontal surfaces, on which reference documents might be placed, up to an average

TABLE 3-1   *Illuminance Criteria Suggestions*

VDT Task	VDT Screen	Frequency of Paper Reference	Paper Document[a]	Users' Ages[b]	Average Illuminance on Paper Task	Ambient Vertical Illuminance[c]	Ambient Horizontal Illuminance[d]
CADD or Conversational VDT	Positive contrast (dark background)	Infrequent	High contrast	Younger	75 lux	75 lux	50 lux
				Older	75 lux	75 lux	50 lux
			Low contrast	Younger	75 lux	75 lux	50 lux
				Older	150 lux	75 lux	100 lux
		Frequent	High contrast	Younger	100 lux	75 lux	75 lux
				Older	200 lux	75 lux	150 lux
			Low contrast	Younger	100 lux	75 lux	75 lux
				Older	200 lux	75 lux	150 lux
	Negative contrast (light background)	Infrequent	High contrast	Younger	100 lux	100 lux	75 lux
				Older	100 lux	100 lux	75 lux
			Low contrast	Younger	100 lux	100 lux	75 lux
				Older	200 lux	150 lux	150 lux
		Frequent	High contrast	Younger	300 lux	200 lux	225 lux
				Older	500 lux	250 lux	375 lux
			Low contrast	Younger	300 lux	200 lux	225 lux
				Older	500 lux	250 lux	375 lux
Data-entry VDT	Positive contrast (dark background)	Frequent	High contrast	Younger	200 lux	150 lux	150 lux
				Older	300 lux	150 lux	225 lux
			Low contrast	Younger	300 lux	150 lux	225 lux
				Older	500 lux	150 lux	375 lux
	Negative contrast (light background)	Frequent	High contrast	Younger	300 lux	200 lux	225 lux
				Older	500 lux	200 lux	375 lux
			Low contrast	Younger	500 lux	200 lux	375 lux
				Older	750 lux	300 lux	565 lux

TABLE 3-1 (continued)

[a]High contrast refers to documents that have inherently good quality contrast—for example, black ink on white paper. Low contrast refers to documents that have inherently poor quality contrast–for example, light gray or light-colored ink on white paper; or black ink on a colored paper.

[b]Typically, older eyes need more light to see any given task equally as well as younger eyes. Yet older eyes are more sensitive to glare and adaptation issues, hence, overlighting of paper tasks can be problematic. "Younger" refers to those eyes typically under 40 years of age; "older" refers to those eyes of 40 years of age or more.

[c]Average vertical illuminance at 1.2 meters above floor throughout room. Suggested values are based on maintaining appropriate luminance ratios as discussed under Luminances, on minimizing veiling reflections from VDT screens, minimizing washout of VDT screens and minimizing adaptation effects. As such, surface reflectances must be in the range of those values reported in Table 3-2. Average-to-minimum ratio should be 2:1 or less. Average-to-maximum ratio should be 1:1.25 or less.

[d]Average horizontal illuminance at work plane height throughout room. Suggested values are based on maintaining appropriate luminance ratios as discussed under Luminances and on minimizing adaptation difficulties. As such, surface reflectances must be in the range of those values reported in Table 3-2. Average-to-minimum ratio should be 2:1 or less. Average-to-maximum ratio should be 1:1.25 or less.

of 565 lux. Note how this differs from the average of 375 lux quoted previously for *conversational* VDT tasks, where the worker makes less frequent reference to paper documents, and so the horizontal illuminance requirements can and should be lower than for data-entry VDT tasks where the worker is likely to refer to paper documents on a frequent basis. If the computer screen is positive contrast for a data-entry task, however, then the surrounding horizontal surfaces should have an average of 375 lux or less on them so that when people glance from the positive-contrast screen to the reference paperwork the brightness of the paperwork is kept to a minimum, thereby reducing transient adaptation effects. This paragraph should be reread carefully to best appreciate the nuances involved in establishing illuminances for various VDT tasks.

Three points of clarification remain that are relevant when seeking these suggested average illuminances: average intensity, uniformity of illuminance, and initial versus maintained illuminance. *Average intensity* refers to the average illuminance level that should not be exceeded. *Uniformity* refers to the degree of variation of light level over a work area or work surface. *Initial versus maintained* refers to the light level present at the initial commissioning of a work space versus the light level present after a period of use.

The average intensity is simply the sum of illuminances at various sample points or positions across the surface or area of interest divided by the total number of points or positions sampled. The average should not exceed the suggested limit, yet could be less than the suggested limit. Looking at it another way, the criteria "average of 225 lux" indicates that the average illuminance should remain at or below the suggested target of 225 lux.

Average illuminance design criteria might be less than the suggested targets if the designer observes that workers' existing conditions are significantly less than suggested maximums. For example, if an existing group of CADD draftspeople are working in a space that is virtually "blacked out," with the exception of some very dim task lights, then it might be culture-shock to develop a new or retrofit environment to meet the average illuminance criteria suggested in Table 3-1. In this situation it is best to meet with users to hear about their perceptions of the work environment as well as to educate the users on appropriate lighting criteria, developing project-specific average illuminance criteria.

Horizontal ambient illuminances must be uniform if disturbing shifts in contrast from one horizontal surface area to another are to be avoided. Such shifts can be distracting and can lead to adaptation problems. A secondary benefit to uniform horizontal ambient illuminance is the ability to set up a workstation at any location and achieve constant and equal lighting conditions.

Uniformity criteria are best cited as ratios of average-to-minimum and average-to-maximum. The smaller these ratios, the more uniform the lighting and the less likely are obvious and annoying contrasts. To avoid apparent dark zones, it might be advisable to limit the absolute minimum to no less than one-half the average (in other words, the average should be no more than two times the absolute minimum). Generally, to avoid glare conditions and overlighting, the absolute maximum probably should be no more than 25 percent greater than the average (in other words, the absolute maximum should be no more than one and one-fourth times the average).

As ratios, then, the average-to-minimum ratio should be less than or equal to 2 to 1 (typically reported as "2:1"), and the average-to-maximum should be less than or equal to 1 to 1 ¼ (1:1.25). Hence, for a workplace with conversational VDT tasks, where the average ambient horizontal illuminance across a work surface is to be designed at 150 lux (based on Table 3-1), the absolute maximum ambient horizontal illuminance at any one point or position on the surface should not be any greater than one and one-fourth times the average, or 188 lux. At the same time, the average should not be any greater than twice the minimum. Since the average is to be 150 lux, the minimum would have to be at least 75 lux.

All of these suggested illuminance criteria are intended to represent in situ maintained values. That is, these illuminances should be maintained over the anticipated life of the lighting system (probably twenty years or so). Additionally, these are the illuminances that should be achieved with furniture, partitions, and any other obstructions in place. Therefore, it is quite likely at the commissioning of an installation that illuminances will be higher than target criteria. Dirt buildup and general wear of lighting equipment and building finishes take several years to accumulate and affect illuminances.

Ambient vertical illuminances that are too high can lead to poor visibility of VDT screens caused by washout of the VDT screen.

In situ (in the actual situation or built environment) conditions include the effects of actual furniture and room partitions, surface reflectances, dirt buildup, and lamp burnout conditions. It is particularly important for the VDT users and the space designers to have a clear understanding of maintenance procedures. Optimal maintenance procedures will result in a more energy-efficient operation and more appropriate visual environment conditions. This typically means that the space designers and users need to meet early in the design process and educate one another on the theoretical and practical aspects of both design and maintenance. See "Maintenance" later in this chapter.

In summary, ambient horizontal illuminance criteria affect task visibility and adaptation and depend on the types of VDT tasks, the kinds of VDT screens, and users' ages. To some extent, preexisting work conditions may also influence ambient horizontal illuminance criteria.

Ambient Vertical Illuminance

Horizontal illuminance, as just discussed, influences adaptation effects. Adaptation effects are also influenced by vertical illuminance, since the eyes' scan around a workplace encompasses both horizontal and vertical surfaces reflecting light. A common problem for people using computers, VDT screen washout is also a result of improper ambient vertical illuminances. In other words, too much light falling onto the VDT screen (primarily a vertical surface) washes out the text/image on the VDT screen. This is particularly problematic with many "low-glare" VDT screens that have rough-textured glass screens or nylon mesh over the screens. These rough surfaces diffusely reflect the light that falls onto them toward the computer users so that veiling images and reflected glare are minimized. The VDT screen exhibits a hazy glow from the diffused reflection of light caused by higher-than-desirable vertical illuminance. This effect is also called veiling reflections—light reflections that are significant enough to veil the text-graphic images.

Ambient vertical illuminance criteria like ambient horizontal illuminance criteria depend on the kind of VDT tasks and the kind of VDT screen. People performing conversational and CADD VDT tasks are more sensitive to vertical illuminance and typically require a lower limit than those performing data-entry VDT tasks. Positive-contrast screens, because of their dark background, are typically more susceptible to vertical illuminance effects and therefore require a lower limit than negative-contrast screens. Since ambient vertical illuminance actually reduces the contrast of the VDT screen, there are also likely to be age sensitivities to vertical illuminance, with younger workers (under 40 years) typically less affected by washout.

For conversational and CADD VDT tasks with positive-contrast screens, the average maintained ambient vertical illuminance at 1.2 meters above finished floor (AFF)—about the height off the floor to the top of a typical VDT screen—should be 75 lux or less. With negative-contrast screens, the average maintained ambient vertical illuminance can be as much as 250 lux, depending on the task conditions and viewers' ages.

For data-entry VDT tasks with positive-contrast screens, the average maintained ambient vertical illuminance at 1.2 meters AFF should be 150 lux or less. With negative-contrast screens, the average maintained vertical illuminance could be as high as 300 lux, depending on the task conditions and observers' ages. Table 3-1 outlines suggested ambient vertical illuminances for a variety of tasks, task conditions, and observers' ages.

Uniformity is an important part of ambient vertical illuminance criteria. To minimize excessive light level swings or variations that could result in excessive contrasts, the average-to-minimum ratio should be 2:1 or less. The average-to-maximum ratio should be 1:1.25 or less.

In summary, ambient vertical illuminance criteria affect adaptation and screen washout. Vertical illuminance criteria depend on the type of VDT tasks, the kind of VDT screen, and users' ages. To some extent, preexisting work conditions may also influence ambient vertical illuminance criteria.

Task Illuminance

The amount of light falling on visual tasks (whether they are on a horizontal surface, vertical surface, or something in between) will influence how well people can see the visual tasks and may influence adaptation effects. Adaptation effects can be deleterious if paper tasks are overlighted, for example. Task illuminance requirements depend on the visual task(s) being illuminated and the users' ages. The VDT screen as commonly experienced today (the screen is internally lighted) needs no external illuminance. Paper documents, however, require some illuminance for them to be visible to workers. Typically, the more detailed and/or the lower the contrast of the paper task, and/or the older the viewers' eyes the more illuminance is required on the paper documents for them to be read with commonly accepted ease, speed, and accuracy. Nevertheless, if paper documents receive too much light, then they can reflect significantly more light than background surfaces and cause adaptation problems that may cause discomfort and/or slow the work process. If overlighted paper documents are viewed in conjunction with a positive-contrast VDT screen, then adaptation problems may be exacerbated. Hence, task illuminance requirements for paper documents not only depend on the characteristics of the paper documents themselves and the ages of the workers but also on the accompanying VDT tasks.

Paper reference documents that may be used infrequently in conjunction with positive-contrast conversational and CADD VDT tasks should have rather low average maintained task illuminances of between 75 and 150 lux. The low end of the range is appropriate if the paper document has inherently high contrast (black ink on white paper and/or large-size text/graphics) and/or if the worker is under 40 years of age. For poorer-contrast paper documents and/or for older workers and/or higher usage, the high end of the range is appropriate. In either instance, the ambient illuminance alone may be sufficient without the addition of task-specific lighting equipment. Care must be taken, however, in recognizing the task plane versus the criteria plane of the ambient illuminance. If the paper reference document is to be used on a paper-document holder, which is advisable, then the preceding task illuminances cited are to be achieved on the paper document mounted to the paper-document handler (which is quite likely near a *vertical* orientation not a horizontal one).

For example, the design criteria for a low-contrast paper task referenced infrequently by observers of 40 years of age or older who are using positive-contrast VDTs for CADD or conversational VDT work is 150 lux. If the reference document is mounted on a paper-document holder, and if the average maintained ambient vertical illuminance is 75 lux, then another 75 lux will be required on just the area or zone of the paper-document holder to achieve the criteria suggestion of 150 lux.

Paper reference documents that may be used infrequently, but in conjunction with negative-contrast conversational VDT and CADD tasks, could have average maintained task illuminances of between 100 and 200 lux. This higher range is acceptable, even desirable, since the VDT screen has a lighter background with dark text/images. Table 3-1 outlines suggested average illuminances on paper tasks for a variety of VDT tasks, task conditions, and observers' ages.

Data-entry VDT tasks by their very nature may involve extensive reading of paper documents. For paper reference documents that are used with positive-contrast data-entry VDT tasks, average maintained task illuminance should range between 200 and 500 lux. The high end of the range should be reserved for poorer-contrast paper documents and/or older workers. For paper reference documents used with negative-contrast data-entry VDT tasks, average maintained task illuminance could range between 300 and 750 lux.

For those data-entry VDT tasks that use dictation recordings as the reference material, viewing of the VDT screen is typically the sole visual task. In this case, ambient horizontal and vertical illuminance criteria will likely suffice as the task illuminance.

Similar to ambient horizontal illuminance requirements, task illuminance requirements should respect some uniformity parameters to avoid adaptation issues and distractions. Average-to-minimum uniformity ratios should not exceed 2:1. Average-to-maximum uniformity ratios should not exceed 1:1.25.

In summary, task illuminance criteria affect adaptation, task visibility, and the extent of visual distractions. Task illuminance criteria depend on the type of VDT tasks, the kind of VDT screen, the kind and frequency of use of reference paper documents, and users' ages. To some extent, preexisting work conditions may also influence task illuminance criteria. Table 3-1 outlines suggested task, ambient horizontal, and ambient vertical illuminance criteria.

Luminances

Although illuminance is largely responsible for the visibility of paper-document tasks, luminance is largely responsible for the visibility of VDT screens, for people's level of brightness adaptation, for perceptions of space, and for glare. It is imperative for the designers and users of spaces to become familiar with the concept of luminance (see a definition in Chapter 1 and a discussion of luminance in Chapter 2) and familiar with luminance quantities. This is best done by experience. Table 2-3 and the discussion on luminance in Chapter 2 illustrate and discuss specific luminances of objects. Luminance is measured brightness. An ordinary piece of white paper has a luminance typically ranging from 40 cd/m^2 to 240 cd/m^2 depending in illuminances. A positive-contrast VDT screen may have a luminance of 6 cd/m^2, while a negative-contrast VDT screen may have a luminance of 80 cd/m^2. The sky outside on a clear day, directly overhead, may have a luminance of 1,000 cd/m^2, and on a cloudy day may have a luminance of 10,000 cd/m^2. In the electronic workplace, the greater the luminance of any one surface or area, the more likely that surface or area will cause one or several of the following:

- ▼ Glare (directly into workers' eyes)
- ▼ Reflected glare (reflected from another surface, like a shiny VDT screen, into workers' eyes)
- ▼ Veiling reflections
- ▼ Veiling images

The uniformity of luminances throughout a workplace is just as important to comfortable, efficient task viewing as are absolute luminances. Luminance uniformity is controlled by using luminance ratio guidelines—the ratio of maximum to minimum luminances. It is suggested that the luminance ratio between any surface or area and any other surface or areas should be 5 to 1 or less. That is, no task, surface, or area should be any more than five times as bright as any other task, surface, or area. To minimize adaptation effects, luminance ratios from one task to another task (e.g., luminance difference between the paper task and the VDT screen) should be 3 to 1. This can contribute to a sense of sameness or blandness throughout a space. To provide for visual interest, however, color contrast should be used rather than luminance contrast.

Luminances and luminance ratios significantly affect how well people can see electronic tasks and how comfortable people are in electronic workplaces. Luminance limits for various surfaces are critical to the success of an electronic workplace. These surfaces include all general room surfaces such as walls and windows, ceiling and skylights, and floor; all work surfaces such as desktops, reference tables, conference tables, and the like; and luminaires, that is, the reflectors, louvers, lenses and housing of the lighting equipment.

Unlike illuminance criteria that are expected to be achieved after several years of operation, luminance criteria should not be exceeded initially. Dirt accumulation factors and lamp lumen depreciation factors are likely to result in decreased luminances over time—and this is desirable. Since luminance can debilitate people's abilities to perform work comfortably and efficiently, then luminance criteria should not be exceeded at the commissioning of a workplace setting. Therefore, unlike illuminance criteria that are reported in terms of maintained values, luminance criteria are initial; that is, luminance criteria should not be exceeded the first day that an electronic workplace is occupied by workers.

Task Surface Luminances

People's concentration on paperwork and/or on VDT screens can be enhanced when the background surfaces are not excessively bright or dreadfully dark. Additionally, when viewing alternatively between paperwork and VDT screen, the adaptation level is affected by the surface luminances of those surfaces immediately surrounding these visual tasks. Such surfaces are desktops, reference tables, work surfaces, and partitions that abut the work surfaces.

To minimize adaptation effects, it is advisable to maintain a 3-to-1 luminance ratio from task to task and from the tasks to the immediately surrounding surfaces. Recognize that a luminance ratio of less than 2 to 1 does not allow for comfortable emphasis on the task itself. Typically, task surface luminances should be in the range of 5 cd/m^2 to 80 cd/m^2—this is quite a wide range, which represents the lower and upper limits.

For people reading paper tasks that have average task illuminances of 75 lux, luminances of nearby surrounding surfaces that are below 5 cd/m^2 are likely to lead to visual fatigue and/or glare as the paper task is likely to be too bright relative to the background surfaces. Luminances of nearby surfaces that exceed 8 cd/m^2 are likely to distract attention from the paper task. Additionally, these relatively bright surfaces may reflect in the user's positive-contrast VDT screen or those of nearby users, causing veiling images and/or veiling reflections.

For people reading paper tasks that have average task illuminances of 750 lux, luminances of nearby surrounding surfaces that are below 50 cd/m^2 are likely to lead to visual fatigue and/or glare, as the paper task and/or the VDT screen are likely to be too bright relative to the background surfaces. Luminances of nearby surfaces that exceed 80 cd/m^2 are likely to distract attention from the paper task. Additionally, these relatively bright surfaces may reflect in the user's VDT screen or those of nearby users, causing veiling images and/or veiling reflections.

General Room Surface Luminances

If wall, floor, or ceiling surfaces are too dark (very low luminances), then people's impressions are likely to be that of a cave—a feeling that one is in a small, confined, and dark space. On the other hand, if walls, floor, or ceiling surfaces are too bright (very high luminances), then people are likely to experience glare and/or difficulty reading VDT tasks. This is especially problematic for positive-contrast VDT tasks.

To limit the possibility of general room surface luminances causing problems, a suggested range of surface luminances is 3 cd/m^2 to 510 cd/m^2. The low end is appropriate for surfaces in spaces where people work on CADD or conversational VDT tasks with infrequent reference to paper documents. The high end is appropriate for window and skylight surfaces in spaces where people work on negative-contrast data-entry VDT tasks with frequent reference to paper documents. In any event, luminance ratios or differences between one surface or area to another should not exceed 5:1. The luminance ratio between a paper task and distant background surfaces (e.g., ceilings, walls) should not exceed 5:1 and should not be less than 1:5. In other words, in situations where there are no windows, luminances should be designed so that the task is not more than five times brighter than any of the distant surface luminances. Where windows or skylights are likely to be used, then the brighter windows and skylights should not exhibit luminances any greater than five times the luminance of the task.

Luminance ratios of up to 10:1 have been reported as acceptable. Experience indicates, however, that such ratios typically lead to luminance "splotches" or "streaks" that are quite noticeable in positive-contrast VDT screens and can be somewhat noticeable in negative-contrast VDT screens.

Luminaire Luminances

Luminaire luminances are most problematic for people using computers and yet are easiest to minimize. Direct luminaires and direct/indirect luminaires exhibit luminances. Even some indirect luminaires have acrylic, glass, perforated metal fins, or slot details that have some luminance.

Luminances of luminaires depend on the angle at which the luminaire is seen. For example, sitting underneath a direct or direct/indirect luminaire and looking up into it typically reveals bright lamp(s). Viewing luminaires at a distance results in much less brightness. Yet the distant luminaires are more likely to reflect from the VDT screen into the user's eyes. Therefore, to minimize the adverse effects of luminaire luminance on VDT viewing, luminance limits vary by angle of view or cutoff. Figure 3-1 illustrates these cutoff angles. At 55°, luminaire luminance should be 850 cd/m^2 or less. At 65°, luminaire luminance should be 340 cd/m^2 or less. At 75°, luminaire luminance should be 170 cd/m^2 or less.

Luminaire luminance information is available from luminaire manufacturers. Testing procedures are such that manufacturers will claim that it is impossible to measure the maximum luminance at a given angle. Nevertheless, instruments and calculations used in developing photometric reports do yield average luminaire luminance at given angles. It is this test information that should be used to determine if a given luminaire will comply with the luminance limit suggestions.

Window Luminances

Window and skylight luminances should follow similar limitations as general room surface luminances while recognizing that windows and skylights are brighter (rather than darker) than the VDT and paper tasks. As such, window and skylight luminances should range from 70 cd/m^2 to 510 cd/m^2. The low end appropriate for areas in which people are working on positive-contrast CADD or conversational

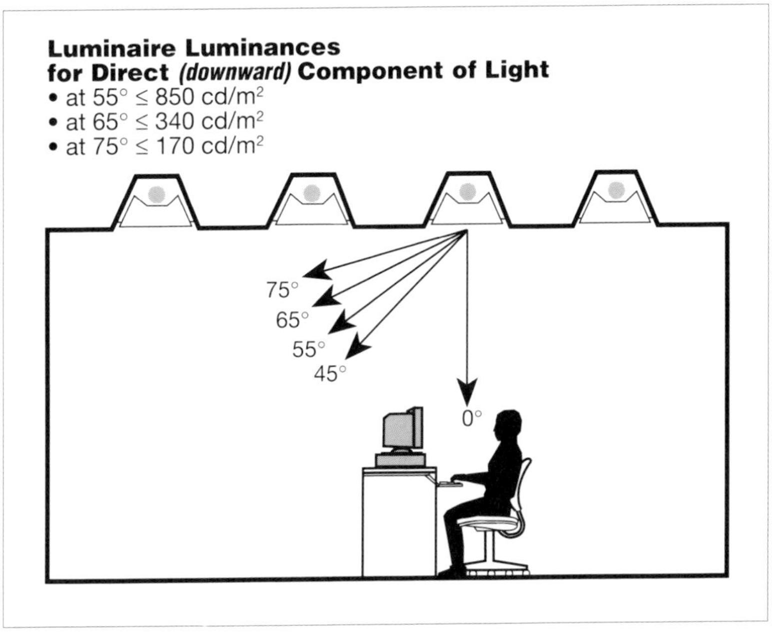

**Luminaire Luminances
for Direct *(downward)* Component of Light**
- at 55° ≤ 850 cd/m²
- at 65° ≤ 340 cd/m²
- at 75° ≤ 170 cd/m²

75°
65°
55°
45°
0°

FIGURE 3-1

*Luminances of direct and direct/indirect lumi-
naires should be limited to these values or less
to minimize veiling images and veiling reflections from VDT screens. This is particularly critical where positive-contrast VDT screens are
used.*

VDT tasks with infrequent reference to paper documents. In such a situation, a skylight or window surface should not exhibit a luminance greater than 70 cd/m², and
less would be preferable. Since typical exterior surface luminances (e.g., sky,
clouds, snow-covered ground, concrete buildings, and so on) can easily approach
10,000 cd/m² (as exemplified in Table 2-3), then the window glazing must be tinted
and/or coated and/or treated with some sort of interior or exterior window covering. Of course, it is reasonable to design the windows and/or skylights for average
conditions and not for extreme brightness conditions, which only occur a small percentage of the year. Nevertheless, it is not unusual for window systems (glazing,
coatings, and/or treatments) to need the capability of providing transmittances as
low as 3 percent, and for skylight systems to need the capability of providing transmittances as low as 1½ percent.

Window coverings can be problematic. Traditional horizontal and vertical blinds
inevitably can be configured so that some people and/or other VDT screens are
exposed to luminances exceeding the suggested limits. A monolithic, image-preserving shading treatment seems to work best (see Figure 2-5). This can minimize
luminances to acceptable levels on most daylight conditions while preserving view
to the exterior—the primary reason for using windows and/or skylights in most
applications. If window covering is to work best most of the time for most of the
occupants, then an automated system of covering is appropriate.

Finally, from a design, engineering, or cost perspective, it may be undesirable
to limit luminance and/or illuminance to the suggested values for all times. For
example, to limit luminance to suggested values on a skylight at all times will mean
that the exterior design conditions must be worst-case. In Detroit, Michigan, USA,
these worst-case conditions for skylights are likely to be on a partly-cloudy day in

June. From a perspective of limiting illuminance, clear-day conditions near noon in June offer the worst-case. The designer must determine the impact on luminance if illuminance criteria are used to establish glazing transmittances.

Surface Reflectances

Surface reflectances have a significant influence on luminances and illuminances. Wild variations in surface reflectances can exacerbate luminance ratio problems, resulting in problems of adaptation, veiling images, and veiling reflections.

Ceiling reflectances should approach 80 percent for best lighting distribution efficiencies throughout the space. Ceiling finishes should be matte, not specular or even semispecular.

Wall reflectances should range between 30 percent and 50 percent. This should include all vertical surface treatments, including window treatments.

Floor reflectances should approach 20 percent. Floor finishes that are darker can contribute to adaptation problems. Additionally, dark floor finishes in conjunction with direct lighting can result in very dark-appearing ceilings, since the dark floor reflects virtually no light onto the ceiling. This exacerbates the cave effect common with low-brightness, direct lighting systems.

Work surface reflectances should range between 20 percent and 40 percent. Darker work surfaces cause luminance ratio problems with paper documents. Higher work surface reflectances can cause luminance ratio problems with VDT screens. Table 3-2 summarizes reflectance recommendations and generic materials that are likely to comply.

Power Budget

With the availability of high-efficiency lighting equipment and with the low-to-moderate illuminances and luminances necessary for people using computers, connected lighting loads should be less than 16 w/m^2 (watts per square meter). Further, occupancy sensors, automated timing devices, computer-based energy management systems, and daylight sensing electric lighting systems can easily result in low-energy-consuming lighting. These strategies can result in effective connected loads of 10 w/m^2 or less.

Subjective Impressions

Previous criteria discussions indicate the significance of proper luminances and luminance ratios. Luminances also appear to be responsible for how people respond to at least the subjective impressions of visual clarity, spaciousness, relaxation, and intimacy. For people using computers, the environment should be pleas-

TABLE 3-2   *Surface Reflectance Guidelines[a]*

Surface	Reflectance (matte)	Complying Materials
Work surfaces	20 to 40%	▼ Light woods ▼ Medium and light laminates ▼ Medium and light ink blotters
Window treatment*	30 to 50%	▼ Medium to light fabrics ▼ Medium to light blinds ▼ Frit pattern glass
Floors	10 to 20%	▼ Medium to light carpet ▼ Medium to light wood ▼ Medium tile
Ceilings	70% and greater	▼ White fabric/cloth ceiling ▼ Pure white mineral tile ceiling ▼ Off-white to white paint on drywall
Walls	30 to 50%	▼ Light fabric ▼ Medium to light vinyl wall paper ▼ Medium to light paint
Open office partitions	20 to 50%	▼ Medium to light fabrics ▼ Medium to light laminates**

[a]Excerpted verbatim with permission from Architectural Lighting Design, *Van Nostrand Reinhold, 1990.*
*Preferably image-preserving to permit view.
**Generally not appropriate acoustically.

ant and comfortable. Feelings of isolation and confinement are not considered appropriate incentives for productivity. Since luminances must be addressed as part of the technical success of viewing computers, then it follows that these luminances should be manipulated in a positive way. For example, Table 3-1 outlines that the design target is 500 lux for average illuminance on a frequently-referenced, high-contrast paper task for an older person performing data entry using a negative-contrast VDT. Since the paper likely has a reflectance of 70 percent, the luminance of the paper is about 111 cd/m^2. Based on the discussion of luminances and luminance ratios, this paper task should not have a luminance any greater than five times any darker-colored wall finishes in the surrounding environment. Therefore, it may be necessary to softly light darker-colored walls to bring luminances into balance, or within the recommended ratio. Wall lighting also enhances the subjective impression of spaciousness. Of course, it may be most appropriate from an energy perspective to lighten the finish of the darker walls that are responsible for the luminance ratio problem. Similarly, the lighter wall finish will enhance the sense of spaciousness.

Color of light may also influence space perceptions and positively influences color perceptions. The newer, triphosphor fluorescent lamps offer a broader spectral distribution of light, thereby rendering skin tones, clothing colors and architectural finishes in a "truer" or "more natural" fashion. Technically, chromatic contrast improves with the triphosphor lamps. In areas where people are using computers, luminance contrast should be minimized. The only other method for introducing visual interest and attraction is to improve chromatic contrast. Lamps with high color rendering and neutral white color (all of the commercially available triphosphor fluorescent lamps) should be considered where people are using computers.

Maintenance

Perhaps first and foremost at the outset of any design project is an establishment of design priorities by the entire design team; this process must include the designers, users, owners, and maintenance personnel. It is poor practice, however, to design any of the environmental systems or architectural elements around the requirements of maintenance. As a general rule, buildings that house people using computers should not be built with the priority to make maintenance easy or nonexistent. Electronic workplaces are built to enable productive operation of people and machines. Inevitably, this means "things" are going to need periodic cleaning, fixing and or replacing.

To make something "maintenance free" or even "maintenance deferred" at the cost of worker comfort and productivity now or later is foolish. This defeats the rationale of building a workplace. To use the "lowest common denominator" components, like standard cool-white fluorescent lamps because these are the only fluorescent lamps available at the local hardware store, is ludicrous. Workplace maintenance isn't an afterthought. Workplace maintenance isn't a "handyman's special" where duct tape, chicken wire, and cheap fittings are used to tide folks over. Workplace maintenance is the maintenance of a place where typically many thousands of dollars an hour are expended on people's salaries, benefits, and work devices (like computers) to achieve the specific end of comfortably and efficiently doing work to make money. Workplaces are like cars, only more important. Periodic maintenance with appropriate and manufacturer-approved parts and service will likely result in longer, more comfortable and more efficient useful life of the infrastructure.

Maintenance is no longer an extension of janitorial services. With the sophistication of energy-saving devices, the sophistication of furniture systems, the sophistication of computers and computer users, and so on, maintenance is one of the many critical functions of the management of facilities' occupants and physical plant. As facility management becomes ever more enthralled with the fine balance of productivity versus overhead costs, lighting will be recognized as less of a hardware-store commodity and more of a carefully crafted, well-balanced, revenue-enhancing system. The challenge to facilities' maintenance folks is bringing maintenance staff up to the level of lighting sophistication, rather than taking lighting down to some level of simple and convenient maintenance procedures.

Maintenance with respect to viewing VDTs has to include at least the maintenance of the VDT screen itself, the lighting system hardware and software (if any), workers, and environmental surfaces (furnishings, walls, ceilings, floors). If any one of these elements is neglected, it is quite possible that task visibility and performance will suffer, that is, productivity suffers.

Maintenance Factors

Maintenance scheduling is important with respect to cleaning and replacement of lighting equipment. Dirt accumulation on room surfaces and luminaires can actually account for about a 5 to 10 percent light loss over a period of several years. Light output drop-off over time (also known as lamp lumen depreciation—LLD) in lamps is a known effect. As lamps are energized, the electricity continuously passing through them degrades some of the lamp components and gases. After several years of typical operation, lamps actually produce 5 to 25 percent less light than at the time of workplace commissioning. The LLD depends on lamp type, with the more efficient fluorescent lamps typically exhibiting a 15 percent light loss over time.

Because the dirt accumulation can be cleaned, and because the lamps can be replaced, these particular losses are known as recoverable light loss factors (RLLF). Of course, the occasional lamp burnout is also considered a recoverable light loss factor. Hence, good periodic maintenance is a necessity for an efficient operation.

Luminaire Cleaning

In most cases, luminaire cleaning can occur every two to three years. This can also be tied to group relamping, so that every other or every third cleaning cycle also includes group relamping.

Luminaire cleaning generally requires more than a ten-second feather-dust. Much of the lighting equipment suited for electronic workplaces uses precisely formed metal and acrylic components for best optical performance. Scratching these surfaces will at least reduce their effectiveness and at most ruin their optical performance. For example, some of today's best direct, parabolic luminaires for electronic workplaces use highly polished sheet aluminum. Using abrasive cleaners and/or wiping against the grain can cause a dulling of the polished sheet, resulting in a haze of light reflections that can cause veiling images reflecting from VDT screens.

Luminaire manufacturers' cleaning instructions should be followed. If unavailable, use nonabrasive cleansers in weak solution and lint-free/nonstatic towels. Soft-ended vacuums can also be used to clean away dust.

Using totally enclosed, lensed lighting equipment to avoid the "cleaning" maintenance is unfounded. Many totally enclosed lensed units are more difficult to clean and actually retain more dirt than opened, louvered luminaires.

Using direct luminaires rather than indirect luminaires to avoid dirt accumulation is inappropriate. Unless the environment is extraordinarily dirt-laden, the dirt buildup difference in indirect versus direct luminaires is minimal. Cleaning indirect luminaires is typically as easy or easier than cleaning direct luminaires. Direct luminaires usually require disengaging the reflector/louver or lens assembly from the housing to facilitate cleaning.

Lamp Replacement

Lamps should be replaced with originally specified lamps or a newer equivalent in terms of physical size, lumen output, color temperature, and color rendering. Disregarding this is a sign of poor maintenance and disrespect for both the space user(s) and the space owner(s).

Lamps should be replaced on a group basis at about 70 percent to 75 percent of rated life. Therefore, if a lamp has a rated life of 20,000 hours, then group relamping should occur after about 15,000 hours of use, which for a typical application might be six years. This group relamping will likely occur before a lot of lamps burn out. Nevertheless, typically it is economically beneficial and energy wise to group replace lamps.

Lamps that burn out before group relamping should always be replaced to maintain appropriate lighting conditions throughout a workplace. This is known as spot relamping. Signs of fluorescent lamp failure include strobing effect, blackening of the ends of the glass tube, very low, flickering output, only a glow at lamp ends, or no operation whatsoever.

Ballast Replacement

Ballasts should be replaced at time of failure. Typically, ballasts will last 40,000 to 50,000 hours of use, which for a typical office application might be fifteen years or so. Ballasts should be replaced with originally specified equipment or with equal performing, energy-improved, flicker-free, noise-free units. Signs of ballast failure for electronic ballasts are no operation whatsoever or rapid on/off sequencing.

Personal Preferences

Although not a quantifiable lighting criterion, it must be recognized that people will have biases on what is most comfortable or "correct" for them. These model guidelines are based on consensus opinion and experience and should be considered a "norm." There are likely to be occupants who will always have some difficulty with the visual environment, whether due to physiological differences with the eye and/or due to previous work environment experiences. Recognize that a program of information dissemination (education) can help people understand what kinds of lighting conditions are best for their task types.

Summary

Prescriptive and/or performance guidelines can be established for lighting electronic workplaces. There are, however, a host of criteria. Table 3-3 summarizes the pertinent criteria and lists the resource tables in this chapter for convenient reference. The lighting designer has many calculational tools available that can be used to predict how lighting systems will perform in given spaces. These tools can be accessed via lighting design consultants, electrical engineers, illuminating engineers, manufacturers' representatives, or purchased by the designer for direct use.

TABLE 3-3A *A Model Lighting Guideline for People Using Positive-Contrast Screens for CADD or Conversational VDT Tasks*

Criteria	Design Targets
Illuminances	▼ See Table 3-1 for Task Criteria ▼ See Table 3-1 for Ambient Vertical Criteria ▼ See Table 3-1 for Ambient Horizontal Criteria
Surface Luminances	▼ **Task surfaces:** 5 to 20 cd/m² ▼ **Room surfaces:** 3 to 200 cd/m² with lower values preferable for intensive and frequent VDT use; and upper values as absolute maximums ▼ **Luminaires:** less than or equal to 850 cd/m² at 55°, less than or equal to 340 cd/m² at 65°, and less than or equal to 170 cd/m² at 75° (see Figure 3-1)
Luminance Ratios	▼ **Task surfaces:** between 2:1 and 3:1 task average luminance to task surface average luminance. ▼ **Room surfaces:** 5:1 any task, surface, object, or area average luminance to any other task, surface, object, or area average luminance
Surface Reflectances	▼ See Table 3-2
Power Budget	≤16 w/m²
Subjective Impressions	▼ **Visual clarity:** consider well-shielded user-controlled task light [typically, uniform central ceiling luminances which best promote clarity cannot be used because of extremely sensitive viewing conditions resulting from the positive-contrast VDT screen] ▼ **Spaciousness:** consider light-colored wall finishes [typically, uniform peripheral luminances which best promote spaciousness cannot be used because of extremely sensitive viewing conditions resulting from the positive-contrast VDT screen]
Miscellaneous	▼ **Maintenance:** establish a maintenance schedule for cleaning luminaires; cleaning VDT screens; replacing lamps; and replacing ballasts ▼ **Personal preference:** develop an education program intended to train employees on the aspects of task position; VDT monitor characteristics and maintenance; eye care, etc.

TABLE 3-3B   *A Model Lighting Guideline for People Using Negative-Contrast Screens for CADD or Conversational VDT Tasks*

Criteria	Design Targets
Illuminances	▼ See Table 3-1 for Task Criteria ▼ See Table 3-1 for Ambient Vertical Criteria ▼ See Table 3-1 for Ambient Horizontal Criteria
Surface Luminances	▼ **Task surfaces:** 7 to 50 cd/m² ▼ **Room surfaces:** 3 to 510 cd/m² with lower values preferable for intensive and frequent VDT use; and upper values as absolute maximums ▼ **Luminaires:** less than or equal to 850 cd/m² at 55°, less than or equal to 340 cd/m² at 65°, and less than or equal to 170 cd/m² at 75° (see Figure 3-1)
Luminance Ratios	▼ **Task surfaces:** between 2:1 and 3:1 task average luminance to task surface average luminance ▼ **Room surfaces:** 5:1 any task, surface, object, or area average luminance to any other task, surface, object, or area average luminance
Surface Reflectances	▼ See Table 3-2
Power Budget	≤16 w/m²
Subjective Impressions	▼ **Visual clarity:** consider well-shielded, user-controlled task light; consider low-level uniform central ceiling luminances (e.g., soft uplighting and/or soft downlighting) ▼ **Spaciousness:** consider light-colored wall finishes; consider low-level uniform peripheral luminances (e.g., soft wall wash lighting and/or very soft wall sconce lighting)
Miscellaneous	▼ **Maintenance:** establish a maintenance schedule for cleaning luminaires; cleaning VDT screens; replacing lamps; and replacing ballasts ▼ **Personal preference:** develop an education program intended to train employees on the aspects of task position; VDT monitor characteristics and maintenance; eye care, etc.

TABLE 3-3C   *A Model Lighting Guideline for People Using Positive-Contrast Screens for Data-entry VDT Tasks*

Criteria	Design Targets
Illuminances	▼ See Table 3-1 for Task Criteria ▼ See Table 3-1 for Ambient Vertical Criteria ▼ See Table 3-1 for Ambient Horizontal Criteria
Surface Luminances	▼ **Task surfaces:** 15 to 50 cd/m² ▼ **Room surfaces:** 8 to 510 cd/m² with lower values preferable for intensive and frequent VDT use; and upper values as absolute maximums ▼ **Luminaires:** less than or equal to 850 cd/m² at 55°, less than or equal to 340 cd/m² at 65°, and less than or equal to 170 cd/m² at 75° (see Figure 3-1)
Luminance Ratios	▼ **Task surfaces:** between 2:1 and 3:1 task average luminance to task surface average luminance ▼ **Room surfaces:** 5:1 any task, surface, object, or area average luminance to any other task, surface, object, or area average luminance
Surface Reflectances	▼ See Table 3-2
Power Budget	≤16 w/m²
Subjective Impressions	▼ **Visual clarity:** consider user-controlled task light; consider uniform central ceiling luminances ▼ **Spaciousness:** consider light-colored wall finishes; consider uniform peripheral luminances
Miscellaneous	▼ **Maintenance:** establish a maintenance schedule for cleaning luminaires; cleaning VDT screens; replacing lamps; and replacing ballasts ▼ **Personal preference:** develop an education program intended to train employees on the aspects of task position; VDT monitor characteristics and maintenance; eye care, etc.

TABLE 3-3D   *A Model Lighting Guideline for People Using Negative-Contrast Screens for Data-entry VDT Tasks*

Criteria	Design Targets
Illuminances	▼ See Table 3-1 for Task Criteria ▼ See Table 3-1 for Ambient Vertical Criteria ▼ See Table 3-1 for Ambient Horizontal Criteria
Surface Luminances	▼ **Task surfaces:** 20 to 75 cd/m^2 ▼ **Room surfaces:** 12 to 765 cd/m^2 with lower values preferable for intensive and frequent VDT use; and upper values as absolute maximums ▼ **Luminaires:** less than or equal to 850 cd/m^2 at 55°, less than or equal to 340 cd/m^2 at 65°, and less than or equal to 170 cd/m^2 at 75° (see Figure 3-1)
Luminance Ratios	▼ **Task surfaces:** between 2:1 and 3:1 task average luminance to task surface average luminance ▼ **Room surfaces:** 5:1 any task, surface, object, or area average luminance to any other task, surface, object, or area average luminance
Surface Reflectances	▼ See Table 3-2
Power Budget	≤16 w/m^2
Subjective Impressions	▼ **Visual clarity:** consider user-controlled task light; consider uniform central ceiling luminances ▼ **Spaciousness:** consider light-colored wall finishes; consider uniform peripheral luminances
Miscellaneous	▼ **Maintenance:** establish a maintenance schedule for cleaning luminaires; cleaning VDT screens; replacing lamps; and replacing ballasts ▼ **Personal preference:** develop an education program intended to train employees on the aspects of task position; VDT monitor characteristics and maintenance; eye care, etc.

4 | Using the Model Guideline: Case Studies

Perhaps the most expeditious method of understanding the nature of the *Model Lighting Guideline* introduced in Chapter 3 is to see the application of the guideline in real situations. The following case studies are from a range of projects in terms of size, cost, and time period. These are based on the author's experiences as a member of Gary Steffy Lighting Design Inc. and respective design teams. With the exception of The

> Many of the case studies presented here served as foundations for developing the Model Guideline in Chapter 3.

MacArthur Foundation Project, these case studies served as foundations for developing the *Model Lighting Guideline* in Chapter 3. The MacArthur Foundation Project occurred after empirical guidelines from other projects were well established. Much of the material previously covered is reiterated here.

On photographic images presented in these case studies, the luminances are exaggerated successively by photography and reproduction technologies. Bright areas are not as bright and dark areas not as dark as they appear in the figures presented here. Contrasts are not nearly as severe as images herein imply. Color characteristics of lamps cannot be conveyed in this black-and-white medium. Colored surfaces appear as black/dark gray on the images here. Without visiting the site, these photos best represent in situ conditions. No photo fill light was used in any of the images shown here. Luminance criteria were achieved to the extent claimed in the write-ups.

CASE STUDY 1

The MacArthur Foundation Headquarters

Project Credits

Project: The MacArthur Foundation Headquarters

Client/Owner: The John D. and Catherine T. MacArthur Foundation

Interior Architect: Powell/Kleinschmidt

Electrical Engineer: McGuire Engineers

Lighting Designer: Gary Steffy Lighting Design Inc.

Photos: Copyright 1992, Robert J. Eovaldi, Courtesy Lithonia Lighting

Award: 1994 IESNA Energy Efficiency in Lighting for Commercial Buildings Award of Excellence

Introduction

The MacArthur Foundation is one of the six largest private foundations in the world. Founded in 1978, the John D. and Catherine T. MacArthur Foundation supports programs in mental health, peace and international cooperation, the environment, the arts, and population. The Foundation owns and is headquartered in Chicago's Marquette Building which was completed in 1894 and designed by Holabird and Roche. With its aggressive support of work on the environment, the Foundation sought to exemplify prudent use of resources in its own facilities. In early 1991, plans were under way for the Foundation to reconfigure existing floors to maximize space use and productivity. At the same time, a lighting review and redesign was commissioned with the goal of reduced lighting energy use significantly.

Existing Conditions

Whether planning new, renovation, or retrofit construction, an appreciation for the existing conditions under which users work is paramount. As discussed previously, a review of existing conditions enables the designer to better develop appropriate lighting criteria for the given tasks; develop a sense of users' expectations of their work environment; and/or determine the extent of necessary education of workers prior to their move to new or retrofitted space. This should at least involve a few-hour review of the existing facility(ies) and should include assessment of visual tasks, task and environmental luminances (and therefore luminance ratios), illuminances, luminaires and lamps, controls, hours of operation, and power budget (watts per square meter).

A lighting survey was undertaken for The MacArthur Foundation. A comprehensive lighting survey should cover the following issues:

▼ daylight availability—windows: their locations/orientations; their treatment; the position of treatment

▼ sky conditions

▼ ambient lighting conditions, including luminaires and their optics, lamps, ballasts, and controls

▼ task lighting conditions

▼ accent/architectural lighting conditions

▼ visual tasks, including reading, writing, facial recognition, video displays, etc (see Table 2-2, Chapter 2)

▼ workstation conditions, including configuration, colors, reflectances

▼ illuminances, including primary horizontal and secondary horizontal, vertical (e.g., VDT screen)

▼ luminances, including work surface(s), task(s), VDT(s), window(s), ceiling, wall(s), luminaire(s), partition(s), partition under binder bin(s), binder bin(s), etc.

▼ controls, including methods (manual, automatic), locations (convenience), and discretization (zoning)

▼ an evaluation of spatial perception—through either surveyors' opinions, and/or one, several, or most of the occupants (at least an informal discussion with several occupants scattered throughout the survey area)

An example of a lighting survey form typical of that used on The MacArthur Foundation Project is shown in Figures 4-1, 4-2, and 4-3. This form, when used in con-

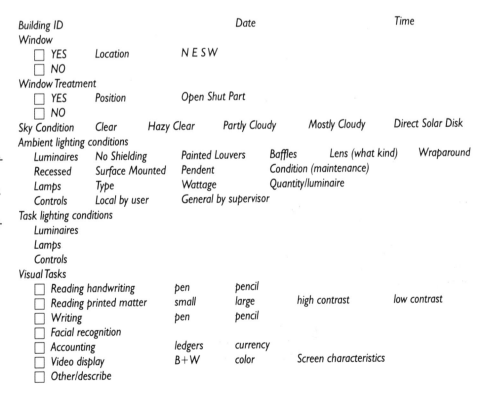

FIGURE 4-1

One portion of the lighting survey form used on The MacArthur Foundation Project. See Figures 4-2 and 4-3 for additional portions of the lighting survey.

Workstation colors
☐ Work surface
☐ Partitions if applicable
☐ Floor
☐ Files
☐ Walls
☐ Ceiling
Comments

FIGURE 4-2

Another portion of the survey used in evaluating existing conditions at The MacArthur Foundation Project. See also Figures 4-1 and 4-3 for other portions of the lighting survey.

Photos

Illuminances
☐ Primary work surface
[insert or sketch a plan layout of work area and mark illuminances]
☐ Secondary work surface   Vertical work surface (VDT)

junction with existing facility architectural reflected ceiling plans and electrical plans, should provide the information just outlined.

From this survey information, a reasonable estimate can be made of existing lighting loads. Additionally, and as or more important, a summary of existing lighting conditions can be developed as a reference base for proceeding with lighting criteria, resolution options, and, ultimately, lighting recommendations.

Prior to complete analysis of the survey information, lighting criteria were established to allow for a reasonable, meaningful analysis of the collected survey data.

Recommended Lighting Criteria

A variety of references, along with experience, are used to establish appropriate lighting criteria. Specifically, documents referenced in the *General References* and *Endnotes* sections served as information and criteria resources. The *Model Lighting*

Luminances
☐ Work surface
☐ Paper task(s)
☐ VDT
☐ Window
☐ Ceiling
☐ Wall(s)
☐ Luminaire(s)
   ☐ N
   ☐ E
   ☐ S
   ☐ W
☐ Partitions
☐ Partition under overheads
☐ Binder bins/overheads
☐ Other

FIGURE 4-3

A portion of the survey used in analyzing the existing conditions at The MacArthur Foundation Project. See Figures 4-1 and 4-2 for additional survey information.

Guideline in Chapter 3 was used. Before establishing criteria, however, a good understanding of the tasks is necessary.

Based on the task survey and an informal interview with the architect and the owner's representative, tasks were established. At the same time, illuminance and luminance measurements were made of existing conditions so that these could be compared to design criteria targets. The Foundation's offices comprised four floors of the Marquette Building. A sampling of private offices on at least three floors was seen as necessary to cover a range of conditions, and to ascertain that significant structural variations did not exist which might impact ceiling heights. Five private offices were surveyed. For similar reasons, and because they occupied a greater amount of space, eight open office workstations were surveyed. Wide variations were sought, including open office workstations near windows, workstations in contiguous layouts and those not (e.g., single workstations located in narrow spaces or small peninsulas created between private offices). The surveys were performed throughout an afternoon with as little fuss as possible so as not to disturb the workers. Nevertheless, if/when the opportunity arose to listen to workers' own thoughts about lighting, notes of these comments were added to the survey.

Visual tasks ranged considerably, although they primarily involved reading. Most of the tasks in the open office areas involved VDTs. Some of these tasks were of conversational nature; others were data (text)-entry-type tasks. In the private offices, although VDTs were used, a majority of the visual tasks were reading of pen and printed text.

Given the anticipated tasks resulting from the task survey, the following recommendations were used as design criteria targets on The MacArthur Foundation Headquarters.

Illuminances

Ambient or general lighting levels should be about 300 lux average maintained at work surface height. Lighting levels for paperwork (the illuminance at the work area) should be no higher than 750 lux average maintained at work surface height and preferably should be about 500 lux to minimize contrast between the paper and VDT tasks. If dark-background (positive-contrast) VDTs are used, then consider the low end of the 500- to 750-lux range for paperwork light levels. If light-background (negative-contrast) VDTs are used, then the high end of the 500- to 750-lux range is acceptable. To maintain visual consistency and avoid transient adaptation effects throughout a space or series of spaces, minimum light levels should be no less than 150 lux average maintained at work surface height. Additionally, to minimize washout of VDT screen images, vertical illuminance (light levels) on screens should be less than 200 lux.

Luminances

Maximum luminance of any surface in the environment should be less than 850 cd/m^2 and preferably 510 cd/m^2. This can be controlled by following both ambient illuminance level and surface reflectance value criteria targets. Additionally, care must be taken to control daylight and to select luminaires that do not exceed the 850 cd/m² limit.

Luminance Ratios

Zero contrast (luminance ratios of 1:1, that is, one surface has the same measured brightness as another, and so on) in a space would provide the best viewing conditions for VDTs. Admittedly, such a space would be quite bland in appearance. Therefore, recognizing the need for some visual variety, it has been reported that the luminance ratios should be 10:1 or less between two different surfaces or two different areas of the same surface (e.g., one ceiling tile to another). In fact, experience shows that luminance ratios of 5:1 or less from area to area or surface to surface—particularly with respect to ceiling, walls, open-plan partitions, and windows—are much more satisfactory in reducing unwanted glare and reflections in VDT screens. That is, the brightest area should be no more than five times the brightness of the dimmest area. This means light-level uniformity is important, as is range in value of surface finishes.

Luminaires

Luminances of direct luminaires should be limited to less than 850, 340, and 170 cd/m² at cutoff angles of 55°, 65°, and 75°, respectively. This will minimize reflected glare images from VDT screens. Figure 3-1 illustrates graphically these suggested luminaire luminance limits.

> Electronic ballasts are noiseless and flickerless and more energy efficient.

Desk-mounted or binder bin-mounted task lights need to be used specifically to light the paper document area and must not be glary. The freestanding desk-mounted lights should be easily adjustable (tilted, moved) by the worker, yet must be shielded so as not to influence nearby workers. Consideration should be given to multilevel switched or dimmable task lighting under binder bins.

Luminaire ballasts offer a significant opportunity for energy savings and for improving the viewing conditions of VDTs by workers. Electronic ballasts are noiseless and flickerless. Also, electronic ballasts operating one or two lamps typically consume about 15 percent less energy than standard energy-saving electromagnetic ballasts.

Surface Reflectances

To attain the luminance and luminance ratio limits previously discussed, surface reflectances should be similar in value from one surface to another. Wild variations in surface reflectance over relatively large areas should be avoided (e.g., a black workstation partition or binder bin and a white work surface). All surfaces need to be matte (see Figure 1-11) to avoid glary reflections. Ceiling reflectance should be between 70 percent and 80 percent. Wall reflectances should be in the range of 30 percent to 50 percent. Floor reflectances should be near 20 percent. Work surface reflectances should be between 20 percent and 40 percent. Where extreme color or value contrast is desired architecturally, surfaces or areas should be highlighted or accented with light to balance darker-surface finishes with those of nearby, non-highlighted lighter-surface finishes. Table 3-2 in Chapter 3 outlines these recommended reflectance guidelines and generic materials/finishes that typically comply.

Power Budget

The ASHRAE/IES 90.1—1989 Standard indicates that for typical open office areas, power budgets should not exceed 20 watts per square meter (w/m^2). Interpretation of the California Nonresidential Standard indicates that the power budget should not exceed 16 w/m^2 for typical office areas (aggregate, including open and private offices). Experience indicates that power budgets as low as 10.8 w/m^2 are achievable in open offices with low to moderate ceiling heights.

Psychological Aspects

Some wall lighting or accent lighting is generally recommended to avoid a gloomy, hazy, bland appearance. Studies done in the mid-1970s at Penn State indicate that such wall lighting can also help to improve impressions of spaciousness. Accent lighting of artwork can help provide distinct distant visual focus points for workers performing extensive close reading or writing work (eye muscle relaxation is enhanced if a distant focus is available for viewing from time to time). Additionally, wall and accent lighting help to promote a less harsh, institutional setting.

The color of light also seems to influence space and color perceptions. The newer, more efficient triphosphor fluorescent lamps offer a spectral power distribution of enhanced green, yellow, and red light, thereby rendering skin tones, clothing colors, and architectural finishes in a "truer" or "more natural" fashion. Experience indicates that perceived truer colors result in an impression of increased brightness. In spaces where warm-toned lamps are used, a softer, less tense, and less formal perception is prevalent. Lamps with high color rendering and warm-toned color should be considered.

Conclusions

A variety of lighting criteria targets, if met, apparently provide a more comfortable, pleasing, and productive work environment. With appropriate surface finishes and lighting equipment, opportunities also existed on The MacArthur Foundation Project for considerably reducing energy resource expenditures. Table 4-1 summarizes the lighting criteria targets specific to The MacArthur Foundation Project.

These criteria were used as a benchmark in the analysis of the information collected during the lighting survey. Also, these criteria were used in the next phase of the lighting study for The MacArthur Foundation—Problem Resolution. Various lighting schemes were developed and reviewed and a recommended preferred scheme was identified.

TABLE 4-1   *Lighting Criteria Design Targets (Recommended for The MacArthur Foundation Offices)*

Criteria	Recommended Design Targets
Illuminances	▼ *Ambient of 300 lux average, maintained at worksurface height; and <200 lux vertical at VDT screen* ▼ *Task of 500 to 750 lux average, maintained on the paper task* ▼ *Minimum of 150 lux average, maintained at worksurface height*
Luminances	▼ **Task surfaces:** *70 to 170 cd/m² for work surfaces* ▼ **Room surfaces:** *510 cd/m² or less for general room surfaces* ▼ **Luminaires:** *less than or equal to 850 cd/m² at 55°, less than or equal to 340 cd/m² at 65°, and less than or equal to 170 cd/m² at 75° (see Figure 3-1)*
Luminance Ratios	*5:1 or less from any task, surface, or areas to any other task, surface, or area (brighter zone to darker zone)*
Surface Reflectances	▼ *Ceiling: ~70%* ▼ *Walls: 30 to 50% (including window treatment)* ▼ *Floor: 20%* ▼ *Work surfaces: 20 to 40%*
Power Budget	*≤16 w/m² aggregate*
Subjective Impressions	*Spaciousness: consider light-colored wall finishes; consider moderate, uniform peripheral luminances*
Miscellaneous	▼ *Ballasts should be inaudible and flicker-free where practical* ▼ *Lamps should be warm to neutral* ▼ *Lamps should be high color rendering* ▼ *Freestanding, desk-mounted task lights need to be optically controlled to eliminate glare* ▼ *Freestanding, desk-mounted task lights should be worker-adjustable in tilt and movement over area* ▼ *Binder bin-mounted task lights should have some multilevel or dimming control* ▼ *Consider subtle wall accenting and/or color accents to avoid gloomy and bland conditions, but follow luminance and luminance ratio limits*

Based on the review of existing lighting conditions and the criteria established, the following conclusions were made regarding The MacArthur Foundation Headquarters lighting that existed prior to retrofit:

▼ Light levels at most task areas were acceptable (with all task lights on).

▼ Light levels on VDT screens in private offices were too high.

▼ Light levels on VDT screens in open offices were acceptable, but this had been achieved by delamping overhead luminaires or switching off overhead luminaires to prevent washout and glare.

▼ Luminances (brightnesses) were too low, particularly on surrounding vertical surfaces.

▼ Luminance ratios, particularly where windows were present, were too high.

▼ Luminaires were inefficient and concentrated too much light downward.

▼ Power budgets were high, given the resulting lighting conditions, and could be significantly improved.

In short, while the premise for the study was "energy reduction," it appeared that energy consumption was not the only lighting criteria to be improved. Energy con-

> ...there appeared to be opportunity to conserve energy **and simultaneously** improve lighting conditions.

servation is necessarily the wise use of available energy resources, and there appeared to be opportunity to conserve energy **and simultaneously** improve lighting conditions. In other words, a double dip in energy reduction was possible—reduce the connected load and increase the chances of improved performance with better quality lighting.

Review of Lighting Alternatives and Recommended Lighting Approach

The "problem resolution" effort included the review of a variety of hardware and design combinations. Generally, however, five basic alternative lighting schemes were reviewed, falling in the following categories:

- ▼ four completely new schemes using state-of-the-art improved color/efficiency lamps, ballasts, luminaires, and control devices
- ▼ one retrofit scheme, retrofitting the existing luminaire with efficient lamps, ballasts, and reflectors/louvers

These five schemes were designed and evaluated on their compliance with the lighting criteria summarized in Table 4-1. A scheme was identified that appeared to best meet all lighting criteria while minimizing energy use.

Before narrowing the design efforts to five schemes, available lamp, ballast, luminaire, and control techniques were identified and reviewed with the Foundation. To ensure that all viable and reasonable energy-efficient technologies were considered, an independent lighting and energy-efficiency consultant was retained to review the analysis. This "second opinion" reinforced the proposed premise of considering both user needs and earth-environment issues.

Alternatives

With relatively low ceilings (2.4 meters up to a maximum of 2.6 meters in all office areas), and due to limited wall area and furniture-mounting options, direct lighting systems were seen as most promising. Nevertheless, one pendent-mounted direct/indirect lighting system was considered to better learn how well such a system might fare under such low ceiling conditions.

Many potential high-quality, low-energy direct lighting systems are available. Limiting alternatives is difficult, but after some experience it became easier to dismiss those systems that would likely fail by a wide margin in at least one of the criteria categories. For example, small-cell, paracube louver luminaires are typically relatively inefficient (not readily meeting efficiency requirements). Using 40-watt, compact fluorescent lamps in large-cell parabolic louvered luminaires generally results in very high luminaire luminances (exceeding criteria). Using large-area luminaires (e.g., 600 mm by 1200 mm or 2 ft by 4 ft) typically yields too much light—both downward intensity, so that vertical illuminances are too high on VDT screens that happen to fall directly underneath, and high-angle luminance, so that luminance limits of 850 cd/m^2 cannot be honored. Five basic fluorescent luminaire types were reviewed:

▼ parabolic 600 mm by 600 mm (or 2 × 2)
▼ parabolic 300 mm by 1200 mm (or 1 × 4)
▼ parabolic 600 mm by 1200 mm retrofit (using existing layout/luminaire chassis) (or 2 × 4)
▼ parabolic 300 mm by 300 mm (or 1 × 1)
▼ 200 mm wide (8-inch), 75 mm deep (3-inch) direct/indirect pendent with bottom suspended 2.2 meters AFF

A variety of lamp and ballast combinations were reviewed within each of these basic luminaire types, including T12, T8, and compact triphosphor fluorescent lamps. Additionally, various available luminaire reflector finishes and configurations were reviewed. This resulted in a review of twenty-two separate lighting systems for each general space type: open office; large private office; and small private office.

Each system was analyzed on its ability to meet the various lighting criteria outlined in Table 4-1, while minimizing energy use. The more promising systems were reviewed in detail, with complete layouts and cost analyses. In each of the five luminaire types, the more promising systems were the following:

▼ parabolic 600 mm × 600 mm with three F17T8 triphosphor lamps and electronic ballasts
▼ parabolic 300 mm × 1200 mm with one F32T8 triphosphor lamp and electronic ballasts (one ballast operating multiple lamps)
▼ parabolic 600 mm × 1200 mm retrofit with two F32T8 triphosphor lamps and electronic ballasts
▼ parabolic 300 mm × 300 mm with two F18BX triphosphor compact lamps and electronic ballasts
▼ 200 mm wide, 75 mm deep direct/indirect pendent with bottom suspended 2.2 m AFF; with on F32T8 triphosphor lamp and electronic ballasts (one ballast operating multiple lamps)

Findings: Open Office Lighting

All of the systems that were reviewed offered annual energy reductions of at least 29 percent from existing conditions. Only one system—the parabolic 300 mm × 1200 mm—met all of the other lighting criteria and offered an annual energy reduction of at least 46 percent (depending on the specific luminaire/lamp/ballast configuration) from the existing conditions. Specifically, the following combination of lighting equipment meets the previously established criteria for The MacArthur Foundation Headquarters:

▼ ambient lighting: a Lithonia Optimax parabolic 300 mm × 1200 mm (1 × 4) system with the GE SP30 triphosphor T8 fluorescent (F32T8/SP30) lamp along with MagneTek/Triad instant-start electronic ballasts (one ballast per two luminaires) for the ambient lighting

▼ architectural fill lighting in circulation areas: a Lithonia Paramax parabolic 300 mm × 300 mm system with the GE SPX30 triphosphor compact lamp (F18BX/SPX30) and MagneTek/Triad instant-start electronic ballasts (one ballast per one luminaire)

▼ architectural fill lighting and task lighting at file banks and circulation areas (where these luminaires double as art accents): a Columbia Parawash 200 mm × 300 mm high-efficiency parabolic-reflector luminaire.

All three of these lighting elements combined yield a connected load of 8 watts per square meter for ambient and architectural fill lighting. An Alkco low-brightness, three-levels-of-light luminaire mounted under binder bins provides task lighting for paperwork (for a total connected load of 12.7 w/m^2 with task lighting included). This system is both earth- and user-friendly. The triphosphor lamp provides excellent color of light to render surface finishes and skin tones "truer."

The parabolic 300 mm × 1200 mm luminaires seen in Figures 4-4 and 4-5, each with one lamp, provide excellent glare control when compared to their multilamp

FIGURE 4-4
Low-brightness of the 1-lamp, 300 × 1200 mm, Optimax parabolic luminaire is readily evident by both the direct view of the luminaire and the view of the VDT screen—no direct or reflected glare and no veiling images or veiling reflections. Courtesy of Robert Eovaldi.

FIGURE 4-5

Lighter-than-normal wall finishes were key to providing an environment that appears appropriately lighted, while minimizing energy consumption. Courtesy of Robert Eovaldi.

counterparts. Recognize that all of the other ambient lighting systems met most criteria, but no other ambient lighting systems met the luminance criteria. Luminance is directly related to glare and VDT-screen reflections. The parabolic 300 mm × 1200 mm, one-lamp luminaire effectively meets this important criterion. Further, when using the Lithonia Optimax, the parabolic 300 mm × 1200 mm provides high efficiency. The electronic ballasts yield significant energy savings while eliminating annoying ballast hum and flicker. Figure 4-6 illustrates the excellent VDT screen vis-

FIGURE 4-6

A photo taken at the same angle as seated eye-height illustrates the high clarity/visibility of the VDT screen under the Foundations's low-energy lighting system. Courtesy of Robert Eovaldi.

ibility under the one-lamp, Optimax parabolic ambient lighting system. Vivid color detail is a result of the GE triphosphor T8 and compact fluorescent lamps, which when combined with electronic ballasts provide the most efficient, white-light, "full-color" lighting available.

Figure 4-5 shows how the 300 mm × 300 mm compact fluorescent luminaire was used to "in-fill" the 300 mm × 1200 mm ambient parabolic lighting.

Findings: Small Private Office Lighting

Again, the parabolic 300 mm × 1200 mm system meets nearly all criteria. Only maximum vertical illuminance is exceeded and this only occurs in one location/orientation. If the VDT is slightly reoriented away from this orientation, providing it is positioned in such a location/orientation, then this criteria limit can be avoided. The parabolic 300 mm × 1200 mm system offers an annual energy reduction of at least 65 percent (at least 81 percent with daylight and occupant sensors) from the existing conditions. This parabolic luminaire uses the same lamping and ballasting as in the open office version discussed earlier, yielding a connected load of 10.8 watts per square meter (for a total connected load of 13.8 w/m^2 including task lighting, where used), with excellent, "truer" color, excellent glare control, no hum, and no flicker.

Findings: Conference Room Lighting

For both demonstrative and visual variety reasons, conference room ceilings were raised and materials changed from the open and closed office ceilings. A variation in lighting was also deemed desirable—to demonstrate other efficient lighting techniques and to introduce visual variety (see Figure 4-7). On the other hand, extreme

FIGURE 4-7
Conference rooms have higher ceilings where direct/indirect lighting is then used to accentuate the ceiling height as well as introduce visual variation from the office areas. Courtesy of Robert Eovaldi.

variation could be detrimental, leading to a "grass-is-greener" syndrome. Therefore, luminance issues were still deemed important. A direct/indirect lighting approach is quite efficient, yet differentiates the conference room ceiling height and the conference setting from the offices. Wall washing was achieved with a T8/electronic-ballasted striplight-in-a-slot approach. Again, comparatively, energy reductions of 61 percent were achieved in conference rooms (at least 78 percent with daylight and occupant sensors). Using the daylight sensor/dimmable ballast has the added feature of providing full-range dimming in the conference room.

Summary

Preretrofit lighting conditions at The MacArthur Foundation accounted for a con-

> The retrofit lighting system meets lighting criteria for people using computers and reduces CO_2 emissions and coal consumption.

nected load of nearly 33.4 w/m² (this includes *all* lighting—ambient lighting, task lighting, fill lighting). Further, the preretrofit lighting conditions at The MacArthur Foundation did not meet guidelines for electronic offices and resulted in delamping, relamping, disconnecting lamps, and adding incandescent task lights. The preretrofit lighting was neither earth-environment-friendly nor particularly user-friendly.

TABLE 4-2   *Recommended Lighting Equipment Summary (for The MacArthur Foundation Offices)*

Application	Luminaire	Lamp	Ballast	Wattage
Open office ambient lighting system	Lithonia Optimax/ PMO-132	1-F32-T8-SP30 (32-watt, 3000°K) triphosphor by GE	MagneTek Triad B132I/232I/332I/432I as necessary to maximize master/slave configuration and minimize connected load	114 watts for 4, 32-watt lamps/ 1 ballast
File/wall wash lighting system	Columbia Parawash/ LWW-1	1-F18-BX-SPX30 (18-watt, 3000°K) compact triphosphor by GE	MagneTek Triad B232I/332I as necessary to maximize master/slave configuration and minimize connected load	51 watts for 3, 18-watt lamps/ 1 ballast
Circulation lighting system	Lithonia Paramax/ PM3-2CF18G-4	2-F18-BX-SPX30 (18-watt, 3000°K) compact triphosphor by GE	MagneTek Triad B232I	35 watts for 2 lamps/1 ballast
Binder bin task lighting system	Alkco/SF332- RSW-HCL-3 DIM	1-F32-T8-SP30 (32-watt, 3000°K) triphosphor by GE	Robertson RS032P	37 watts at full power for 1 lamp/ 1 ballast
Private office occupancy sensor	Lithonia LUS-Series/Equinox	N/A	N/A	N/A
Private office daylight sensor	Lithonia LEQ/ DPC/Equinox	N/A	N/A	N/A
Private office ambient lighting system	Lithonia Optimax/PMO-132	1-F32-T8-SP30 (32-watt, 3000°K) triphosphor by GE	Advance Mark VII (dimmable)	60 watts for 2 lamps/1 ballast
Conference room pendents	Neoray 66DIP/ 1-8/S79	1-F32-T8-SP30 (32-watt, 3000°K) triphosphor by GE in cross-section	Advance Mark VII (dimmable)	60 watts for 2 lamps/1 ballast

The lighting criteria guidelines set forth in Table 4-1 were used to develop a new lighting system. Lighter-than-normal finishes were used to comply with criteria. Colorful finishes in conjunction with high-color-rendering lamps provide visual interest without harsh contrasts. Further, given the low-ceiling-height constraints, a VDT-sensitive, high-efficiency, parabolic, 300 mm × 1200 mm, one-lamp lighting system was used in conjunction with high-efficiency task and fill lighting.

All lamps are the high-efficiency, T8 or compact, triphosphor fluorescent type operating on high-power factor electronic ballasts. Where low-wattage fluorescent lamps are used in task lighting and fill lighting, high-power-factor electomagnetic ballasts are used, if electronic ballasts are not available.

Table 4-2 summarizes the recommended lighting equipment for The MacArthur Foundation Headquarters. Implementing these recommendations led to significant energy reductions (in excess of 61 percent overall) from preretrofit conditions *and* provided a lighting system that meets state-of-the-art electronic workplace lighting guidelines. The result is an earth-environment- and user-friendly work setting with an aggregate total "raw" connected lighting load of 12.9 watts per square meter, including all task, ambient, fill, and art lighting. In other words, over just 3,100 square meters of retrofitted space, the Foundation will reduce carbon dioxide emissions by 124 metric tons a year and will save 26 metric tons of coal a year while simultaneously improving the office environment!

CASE STUDY 2
The Prudential Network Command Center

Project Credits
Project: The Prudential Network Command Center
Client/Owner: The Prudential
Interior Architect: The Planned Expansion Group
Electrical Engineer: Flack + Kurtz
Lighting Designer: Gary Steffy Lighting Design Inc.
Photos: Copyright 1991, Robert J. Eovaldi, Courtesy Peerless Lighting

Introduction

The Prudential Insurance Company of America is the largest insurer in the United States. To maintain close scrutiny of electronic operations at many of its own corporate facilities around the country, a central monitoring facility had been established a number of years ago. With the proliferation of electronic media, this facility was renovated/expanded in 1988. The existing facility was not surveyed as part of the lighting design assignment. As such, it was incumbent on the interior designers and lighting designers to learn the layout of work areas and to understand the kinds of visual work to be undertaken and the intended use of the facility. Since the existing facility was not surveyed, it was most important to establish definitive lighting design goals and design toward these goals. Further, these goals had to be based on some sort of industry-accepted guidelines or norms, rather than on the experience of existing conditions in conjunction with published guidelines, which was the case in The MacArthur Foundation (see *Case Study Number One*).

Recommended Lighting Criteria

At the time of this project's inception (late 1987/early 1988), useful published guidelines on lighting for VDTs were limited to an early version of the German DIN Standard, the Chartered Institute of Building Services Engineers *Technical Memoranda TM6: Lighting for Visual Display Units*,[21] and the International Commission on Illumination *Vision and the Visual Display Unit Work Station* (CIE-60).[20] Practical experience over the early portion of the 1980s showed that not limiting ceiling luminances and luminance ratios typically resulted in environments which users found unsatisfactory for performing electronic tasks.

Given the program information collected by the interior designers and the client, it was quite apparent that the Network Command Center (NCC) would

> The Network Command Center (NCC) requires continuous and at times arduous performance of electronic tasks.

require continuous and at times arduous performance of electronic tasks. Each worker could use two or three VDTs at any one time as well as view one or two large-scale, front-projected images. Workers could be contacted by phone from a computer user at a remote site and based on the phone conversation could input/extract information from the array of VDTs available. Although some reading/writing of hard copy could occur, primary tasks were viewing VDT screens and projection screens. This is a classic example of a conversational VDT task.

The center is operated twenty-four hours a day, seven days a week, year-round. This necessitates that considerable thought be given to the issues of maintenance and the temporal quality of the lighting. Maintenance is particularly critical because of the necessity of continued operations. Lamp replacement should be relatively easy and infrequent; hence, lamps should be easily accessible and of long life. Temporal quality of light is important because of the shift nature of the work. Different shifts and operators will likely prefer different luminances; hence, lighting changes over time should be considered.

Illuminances

Performance of paper tasks tends to be infrequent. Many of the VDT screens can display information in a negative-contrast fashion, although some information, because of its nature, is displayed in positive-contrast fashion. With the conversational/interactive nature of the visual work, ambient lighting levels on work surfaces could range from 75 lux (for positive-contrast VDT use) up to 150 lux (for negative-contrast VDT use).

Luminances

At the time this project was undertaken, maximum luminance values of room surfaces were not considered a criterion. No quantification of maximum luminance had been made in enough settings. Hence, one way of minimizing luminance as an issue was to avoid any luminance or harsh light directly from luminaires; that is, using lighting hardware that exhibited little or no luminance. This led to an immediate narrowing of hardware to indirect lighting equipment. Illuminances and luminance ratios were used as guides in addition to seeking lighting hardware that exhibited very little luminance. The center has no daylight-admitting media.

Luminance Ratios

Luminance ratios at the time of this project's undertaking were primarily applied to ceilings, since it was observed that the veiling reflections and veiling images were in large part a result of the VDT screen "seeing" the ceiling plane. Experience had shown that ceiling luminance ratios should not exceed 10:1 (maximum luminance to minimum luminance) and that 5:1 was preferable.

Luminaire Luminances

As cited under *Luminances*, luminance issues were addressed in a qualitative fashion—by eliminating anticipated sources of high luminance. Indirect lighting was selected as the concept design approach for general or ambient lighting, thereby eliminating any direct lamp luminance issues. Nevertheless, if ceiling luminance uniformity was not upheld, indirect lighting could still provide an unfavorable work setting for continuous viewing of VDTs.

Surface Reflectances

Without rigorous environmental luminance and luminance ratio criteria the issue of reflectances became more of a localized design issue. For example, where wall surfaces would be used as backdrop to overhead projection screens, lower wall reflectances would help minimize projection screen washout and enhance attention onto the projection screen. However, though this is true, it only works for short-term viewing conditions like those encountered at the center. For longer-term viewing conditions, the darker background can cause adaptation effects discussed previously, hence a lighter background would be preferable.

To help mitigate the "cave effect" of low-reflectance, unlit walls, it was decided that wall lighting of some sort should be used. This decision was reinforced by the review of *Psychological Aspects*, discussed later.

Power Budget

Power budget was not used as a criterion. Given the size of the space and the criticality of its use, power budget was not considered as a primary goal. Additionally, as lighting criteria were developed, it became clear that the lighting load would remain relatively small.

Psychological Aspects

Two issues related to subjective impressions were deemed worthy of consideration. First, the lighting criteria could lead to a relatively bland visual space. As such, some visual interest was desired, perhaps from the lighting equipment or from lighting effects. Second, because of the nature of the work, a lighting approach that provided some sense of relaxation was desirable. Because uniform overhead lighting typically promotes a relatively tense setting, and peripheral wall lighting typically promotes a more-relaxed setting, it was decided that some peripheral wall lighting would be used in conjunction with the ambient lighting.

Recommended Lighting Approach

A combination of a linear, widespread, symmetric-distribution indirect fluorescent pendent-mounted LST-series lighting system by Peerless Lighting was used, and an asymmetric, bulkhead-mounted version of the same luminaire provided ambient lighting in the space. Lamping was the then-new triphosphor T8 lamps (Sylvania Octron) for improved color, efficiency, and light distribution. The widespread distribution pendent version along with the asymmetric bulkhead-mount version were necessary to maintain luminance ratios on the ceiling while minimizing suspension length. Pendent version luminaires are 200 mm in width by 120 mm in height by 3.7 meters in length and are suspended 400 mm below the ceiling (to top of luminaire) and spaced on 1.8 meter centers. The bulkhead-mounted asymmetric luminaires are mounted continuously around the coffer edge (see Figure 4-8). Ceiling luminance ratio is 4.7:1. Although maximum average illuminance on the work surface could reach 400 lux, a dimming system was used to limit the amount of ambient light and to provide a few user-selected scenes, which all remain within recommended criteria. To achieve recommended ceiling luminance ratios within the architectural configuration shown in Figure 4-8, resulting luminaire spacings could provide 400 lux on the work surface. Figure 4-8 shows ceiling uniformity and resulting high visibility of VDT screens. Additionally, Figure 4-8 shows the good visibility of overhead projection screens.

To provide user flexibility according to shift and task, the ambient lighting is dimmable with Lutron Hi-Lume dimming ballasts, resulting in a dimming range from 1 percent to 100 percent with no flicker or hum. A preset scene control system was used to enable users to change luminances and lighting levels within prescribed limits but not allow full-range dimming that could lead to adaptation effects and

> Because peripheral wall lighting typically promotes a more-relaxed setting, it was decided that some peripheral wall lighting would be used.

FIGURE 4-8

Widespread indirect ambient lighting provides ceiling luminance ratios less than 5:1, maintains an apparent brightness to the space, eliminates the "cave effect," and controls luminances on walls to a uniform, low level for good projection screen visibility. Courtesy of Robert Eovaldi.

veiling reflections. The work surface illuminance taken during the photo shoot for Figure 4-8 shows that users typically choose between 50 and 75 lux.

Wall sconces manufactured by Ron Rezek (now a part of Artemide) were used to provide some luminance to the low-reflectance walls, to introduce a more relaxed sense to the setting, and to provide some visual interest that might otherwise be lacking where uniformity issues are important. Incandescent lamps were used for ease of dimming. Table 4-3 summarizes the recommended lighting equipment for The Prudential Network Command Center

TABLE 4-3   *Recommended Lighting Equipment Summary (for The Prudential Network Command Center)*

Application	Luminaire	Lamp	Ballast	Wattage
Ambient lighting system	*Peerless Lighting LST-010001*	*1-FO32/735 (32-watt, 3500°K) triphosphor by Sylvania*	*Lutron Hi-Lume*	*67 watts for 2, 32-watt lamps/ 1 ballast*
Wall sconce lighting system	*Ron Rezek Sconce 411*	*2-75A19*	*N/A*	*150 watts for 2 lamps*

CASE STUDY 3

Steelcase Inc. Corporate Headquarters/1994

Project Credits

Project: Steelcase Inc. Corporate Headquarters
Client/Owner: Steelcase Inc.
Interior Architect: Steelcase Corporate Design Group
Electrical Engineer: Electrical Management and Design
Lighting Designer: Gary Steffy Lighting Design Inc.
Photos: Copyright 1994, Robert J. Eovaldi, Courtesy Peerless Lighting

Introduction

Steelcase Inc. is the world's largest producer of office furnishings. With its beginnings in 1912, Steelcase has a long history of developing furnishings for office workers and an extraordinary understanding of the relationship between physical work environment, its arrangement, and workers' comfort, satisfaction, and productivity. In 1983, a 500,000-square-foot headquarters was completed for Steelcase by The WBDC Group. At the time, this facility was a microcosm of architectural-, engineering-, interiors-, and furnishings-leading-edge technologies and applications. Even so, by 1993, new technologies and continued demands in energy use and its costs prompted Steelcase to revisit Corporate Headquarters and develop a model interior for the next five to ten years.

Existing Conditions

A formal lighting survey was not undertaken. Gary Steffy Lighting Design had provided the second-phase lighting consultation on the 1983 project and was involved in other projects where annual and semiannual visits to Corporate Headquarters allowed a firsthand ongoing review of tasks, work group arrangements, and maintained lighting conditions. Every workstation has at least one VDT, and nearly all have the capability to use full-color presentation. Most people use the VDT at least part of the day; however, the area that was retrofitted included a travel agency, where people use VDTs on a conversational basis. In another portion of the area that was retrofitted, a legal group was accommodated. Here, not only are VDTs used, but also a reasonable amount of reading of hard copy, as well as meeting with other people, occurs.

Recommended Lighting Criteria

The lighting criteria outlined for this project and used to develop design solutions were identical to those used for The MacArthur Foundation Headquarters. Although intensive VDT use was expected to occur, there was also the likelihood of periods of intensive reading of hard copy and/or meeting with other people.

Illuminances

Ambient or general lighting levels should be about 300 lux average maintained at work surface height. Where paperwork occurs, lighting levels should not exceed 750 lux and preferably should approach 500 lux to minimize contrast and resulting transient adaptation when viewing between paper documents and VDTs. To avoid transient adaptation effects that might occur when viewing from one surface or area to another surface or area, the minimum illuminance should be about 200 lux at work surface height.

To limit the possibility of VDT screen washout, vertical illuminances at 1.2–meters AFF in the four primary viewing directions (e.g., north, south, east, and west) should be less than 200 lux.

Luminances

Maximum luminance of any surface in the environment should be less than 850 cd/m^2 and preferably 510 cd/m^2.

Luminance Ratios

Luminance ratios of 5:1 (maximum to minimum) or less from area to area or surface to surface are necessary if significant reduction and/or elimination of unwanted glare and reflections in VDT screens are desired. This suggests that light-level uniformity from one area or surface to another is important. Similarly, surface reflectances must not vary wildly from one another.

Luminaires

Luminaire luminances should not exceed 850 cd/m^2 at 55°, 340 cd/m^2 at 65°, or 170 cd/m^2 at 75° (see Figure 3-1).

Desk-mounted and/or binder bin-mounted task lights should provide the user with some sort of user control, either physical adjustment in height, orientation or location over the task surface, and/or light level control via dimmer or multilevel switch mounted on the luminaire.

Electronic ballasts should be used to avoid annoying hum and flicker, as well as to maximize efficient operation of the lighting system.

Surface Reflectances

As indicated previously, wild variations in surface reflectances should be avoided. All surfaces should have a matte finish to limit glary reflections. Ceiling reflectance should be near 80 percent. Wall reflectances should range between 30 percent and 50 percent. This includes the partial-height workstation partitions. Floor reflectances should be near 20 percent. Work surfaces should have reflectances ranging between

20 percent and 40 percent. Table 3-2 in Chapter 3 should be referenced for additional surface reflectance information.

Power Budget

Although the ASHRAE/IES 90.1—1989 Standard indicates that 20 w/m^2 is acceptable, given previous experiences (see The MacArthur Foundation Headquarters case study), a power budget of no more than 16 w/m^2 is established as a design target.

Psychological Aspects

To achieve the luminances and luminance ratios discussed previously, many times the environmental setting is visually bland. To minimize this bland sense, some accent lighting is desirable. Additionally, wall lighting or vertical accents help provide a relatively more spacious and relaxed setting.

Improved chromatic (color) contrast seems to result in an impression of increased brightness. The newer, more efficient triphosphor fluorescent lamps offer improved color rendering and thus provide an impression of increased brightness. Warm-toned lamps seem to provide a softer, less tense setting, perhaps enhancing more interpersonal relationships or interactions. Lamps with high color rendering and warm-toned color should be considered.

Conclusions

The lighting criteria used for developing the retrofit lighting at Steelcase Corporate Headquarters were identical to those criteria used for The MacArthur Foundation Headquarters, and are reported in Table 4-1. The lighting conditions of the existing spaces are best exemplified in an article, "A Maker of Office Furniture Develops Lighting and Acoustics for Its Own Office," appearing in the November 1983 issue of *Architectural Record*. Figure 4-9 illustrates the lighting conditions of a typical open office area shortly after the project was originally commissioned in 1983. Existing lighting conditions, prior to retrofit, are summarized as follows:

▼ Ambient light levels were generally appropriate at about 300 lux.
▼ Task light levels at many task areas were somewhat high with all task lights energized (light levels at task areas approaching 900 lux with all task lights energized in conjunction with the ambient lighting). Articulated movement of task lighting equipment was limited and this was considered a problem.
▼ Light levels on VDT screens in private offices were quite low and therefore acceptable.
▼ Light levels on VDT screens in open offices were acceptable.
▼ Luminances were too high in some areas, particularly on ceiling surfaces above HID uplights.
▼ Luminance ratios were too high (in some cases approaching 50:1).

FIGURE 4-9

Conditions prior to retrofit are shown here, with ceiling luminances and luminance ratios that exceed current-practice guidelines. Courtesy of Steelcase Inc.

▼ Luminaires were inefficient and concentrated too much light directly upward, not allowing for a soft, uniform spread of light across the ceiling plane.

▼ Power budgets were reasonable (20.5 w/m^2)

In short, a number of lighting criteria were not within current-day guidelines and, if improved, would likely result in a more comfortable and productive work setting.

Review of Lighting Alternatives and Recommended Lighting Approach

Client representatives indicated that the consensus was to pursue indirect lighting alternatives for the general or ambient lighting retrofit. Previous experience by Steelcase personnel in a variety of other work settings using indirect lighting led to a general agreement that this approach is most desirable when ceiling heights allow an appropriate and economically viable layout to be developed.

A variety of completely indirect approaches were considered for the ambient lighting. With each approach, however, several accent or architectural lighting approaches were investigated to address the issue of "blandness" often associated with uniform general lighting. The indirect ambient lighting systems included those that are continuous and linear and that use 900 mm, 1200 mm, and/or 1500 mm lamps. To keep the scale of the indirect equipment small, which was felt more appropriate for human occupants, the smaller T8 and high-wattage compact fluorescent lamps were considered for use in luminaires using high-efficiency, widespread-distribution optics to spread the light across the ceiling uniformly even with relatively great center-to-center spacings. Accent options included adjustable recessed units or surface-mounted monopoint units for art highlighting or recessed units for washing entire wall surfaces rather uniformly. Care had to be taken with the accent lights to either minimize their use in areas where VDT use would be continuous or limit their output so that luminaire luminance and surface luminance ratios would not be exceeded.

Findings: Open Office Lighting

The following combination of lighting equipment meets or exceeds the previously described criteria for the Steelcase Corporate Headquarters:

▼ ambient lighting: a Peerless Rounded Softshine uplight suspended 400 mm below ceiling (clear) and of continuous length (2.4 m, 3.6 m, 6 m, and 7.2 m were the various lengths used depending on architectural conditions) by 175 mm in width and 75 mm in height with one lamp in cross section (the GE SP30 triphosphor T8 fluorescent F32T8/SP30) along with Motorola rapid-start electronic ballasts (one ballast serving two lamps where possible)

▼ architectural fill lighting in circulation areas: a Louis Poulsen semi-recessed downlight of 130 mm diameter with a luminous white cone projecting 100 mm below the ceiling; using one 13-watt compact fluorescent lamp (a GE F13DBX23T4/SPX27) along with a high-power-factor electromagnetic ballast

▼ task lighting: a Steelcase task light under binder bins with T8 lamping and standard high-power-factor electromagnetic ballasting along with an on/off switch.

These lighting elements combined to yield a connected load of approximately 12.2 w/m^2. Figures 4-10, 4-11, and 4-12 illustrate conditions in the open plan offices after the retrofit.

The 2.5 meters of lamps nearest the window wall (not shown)—which amounts to about 25% of the ambient lights—are controlled by Advance Mark VII ballasts and operated by a Lutron Micro-Watt photo sensor to take advantage of the daylight. Here, a lower-limit setting on the dimmable uplights is advisable in order to prevent the obvious visual cue of lights being switched off which might either interfere with workers' concentration or perhaps miscue workers that there is an electrical problem. Additionally, all lights in the open plan are on time and occupancy sensor control. This allows a complete "sweep" of the lighting at predetermined times (e.g., lunch hour, 6:00 P.M., etc.), yet the occupancy sensor will energize lights in small zones if occupants are sensed in that zone. The occupancy sensors and zoning of the lights alone can provide

FIGURE 4-10

Single-lamp cross-sectional indirect Softshine lumi-naires provide excellent ceiling luminance uniformity with a clear mounting distance of 400 mm between ceiling and top of luminaire. Ceiling height is 2.84 m. Courtesy of Robert Eovaldi.

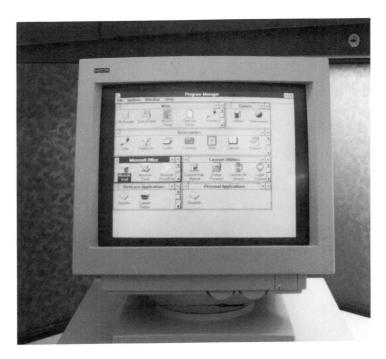

FIGURE 4-11

A close-up of the VDT monitor in Figure 4-10 shows that appropriate luminaire and room surface lumi-nances and luminance ratios combined with nega-tive-contrast screens result in excellent screen view-ing conditions. Courtesy of Robert Eovaldi.

an energy use reduction from 10 percent to 25 percent. The photosensor controls are expected to provide an additional energy reduction of at least 10 percent.

In other words, with the controls in operation, it is likely that the equivalent connected load will be reduced from 12.2 w/m² to 9.8 w/m² or less.

Findings: Private Office Lighting

A full-height glass partition with a vision-glass sidelight and clerestory separates private offices from the open plan. This helps give a more open appearance to the offices and provides a visual connection to both the surrounding office area as well as to the exterior. This vision glass clerestory allows workers in both the private offices and in the open office area to experience the sense of a larger work environment—since the perception is that the ceiling continues for a good 3 to 3.5 meters beyond the actual private office wall. This portion of the ceiling is also "seen" by VDT monitors that might be facing in the direction of the private offices. Therefore, this ceiling surface needs to meet the same luminance and luminance ratio criteria as the ceiling in the open office. At the same time, however, if the private office is empty, using electricity to operate lighting equipment is undesirable.

> With occupancy sensors and photo sensors the effective connected lighting load in the open office area is less than 10 w/m².

Indirect lighting was chosen to maintain similar ceiling luminances and luminance ratios as achieved in the open office. Wall-mounted indirect Peerless Rounded Softshine Asymmetric luminaires, with single-lamp cross section using GE

FIGURE 4-12
The architectural accent lights along the circulation path help alleviate blandness associated with the uniform brightnesses necessary for electronic offices. These accents also provide a visual cue identifying circulation zones for visitors. Courtesy of Robert Eovaldi.

F40BX/SPX30 lamps and Lutron Hi-Lume ballasts, were used to provide ambient lighting. This ambient lighting was tied to an occupancy sensor and set at a low-end setting of 40 percent so that even when the office was unoccupied a soft general glow would be cast onto the ceiling to prevent a "black-hole" effect and to prevent low luminances and excessive luminance ratios.

Task lighting in the private offices was achieved with the Steelcase task light under binder-bins. This occurred over a credenza-like worksurface. At the main desk work surface, task lighting was achieved with Kurt Versen compact fluorescent downlights. Each downlight uses a single GE F18DBXT4/SPX27 lamp (an 18-watt lamp) with a high-power-factor electromagnetic ballast for a connected load of about 25 watts per each luminaire. Lamps are oriented vertically in the luminaire, which allows for a smaller ceiling aperture (about 150 mm) and a lighted appearance very similar to that of an incandescent downlight. Total connected load in the private offices was about 37 w/m^2 with an equivalent connected load of about 27 w/m^2 due to occupancy rates and occupancy sensor benefits.

Findings: Conference Room Lighting

Most of the conferencing space is located along the east window wall. Similar to the private offices, clerestories are used to provide a connection to the exterior for those workers in adjacent office areas. To provide a brightness transition from window wall to the nearby open plan ceilings, indirect Peerless Rounded Softshine pendent luminaires are used. This also provides an ambient light level sufficient for brief, conversation-only meetings. For longer meetings and for those meetings involving the reading of paperwork, Kurt Versen compact fluorescent downlights are again used for emphasis lighting on the table work surface. Kurt Versen incandescent adjustable accent luminaires are used with GE 50PAR30/HIR/NFL lamps (50-watt, halogen infrared narrow flood, PAR30) to provide some visual relief and to balance window wall and task luminances. Total connected load is about 22 w/m^2.

Summary

Preretrofit lighting conditions at Steelcase Corporate Headquarters 2-East/South Area accounted for a connected load of 20.5 w/m^2, which isn't too far out of line. Nevertheless, lighting criteria such as luminance and luminance ratios were not within today's standards and energy savings could be achieved with the lighting.

With the ceiling height of 2.84 m, and the client's desire to consider indirect lighting approaches for ambient lighting, a high-efficiency, widespread-distribution indirect pendent luminaire was recommended for ambient lighting. Where accents and task lighting were required, efficient triphosphor fluorescent technologies were recommended, including the T8 lamps and compact fluorescent lamps (both high- and low-wattage types). Where practical from a purchasing/installation schedule perspective, electronic ballasts were recommended to optimize efficiencies while eliminating annoying flicker and hum. For the few areas requiring specific spot-

TABLE 4-4   *Recommended Lighting Equipment Summary (for The Steelcase Corporate Headquarters Retrofit)*

Application	Luminaire	Lamp	Ballast	Wattage
Open office ambient lighting system	Peerless LD7-010450	1-F32-T8-SP30 (32-watt, 3000°K) triphosphor by GE in cross section	Motorola M1-RN-T8-1LL-120 or M2-RN-T8-1LL-120 (electronic) as necessary to maximize master/slave configuration and minimize connected load	60 watts for 2, 32-watt lamps/1 ballast
Open office ambient lighting system with daylight sensor (near window walls)	Peerless LD7-010450	1-F32-T8-SP30 (32-watt, 3000°K) triphosphor by GE in cross section	Advance Mark VII RDC-2S32-TP (dimmable electronic with range from 20% to 100%) to maximize master/slave configuration and minimize connected load in conjunction with a Lutron Micro-Watt photocell and controller	61 watts for 2, 32-watt lamps/1 ballast when lights are fully energized and 15 watts at 30% full output (note: lights are never switched off even with high-level daylighting, since subjective reactions by workers to "lights out" is negative)
Circulation lighting system	Louis Poulsen PUL-R-225-White/120	1-F13-DBX23T4-SPX 27 (13-watt, 2700°K) double compact triphosphor by GE	Poulsen-supplied high-power-factor electromagnetic	16 watts for 1 lamp/1 ballast
Binder bin task lighting system (throughout open plan and private offices)	Steelcase T8	1-F32-T8-SP30 (32-watt, 3000°K) triphosphor by GE or 1-F25T8-SP30 (25-watt, 3000°K) triphosphor by GE depending on size of binder bin and work surface	Steelcase-supplied high-power-factor electromagnetic	37 watts for 1-F32T8 lamp/1 ballast or 33 watts for 1-F25T8 lamp/1 ballast
Open plan and private office occupancy sensor	Watt Stopper Ultrasonic Sensors	N/A	▼ Open Plan Controller: Honeywell ▼ Private Office Controller: Lutron Micro-Watt	N/A
Private office ambient lighting system	Peerless LD8-010452	1-F40/30BX-SPX30/R S (40-watt, 3000°K) high-wattage compact triphosphor by GE in cross section	Lutron Hi-Lume FDB-2227-120-2 (dimmable electronic with range from 5% to 100% with the specified lamp)	73 watts for 2 lamps/1 ballast
Private office task lighting system (oriented over worksurface)	Kurt Versen P625-HP-W (Wheat cone for best incandescent-like appearance of compact fluorescent downlight)	1-F18DBXT4-SPX27 (18-watt, 2700°K) double compact triphosphor by GE operated in vertical orientation for best incandescent-like appearance of downlight)	Kurt Versen-supplied high power factor electromagnetic	21 watts for 1 lamp/1 ballast

TABLE 4-4 (continued)

Application	Luminaire	Lamp	Ballast	Wattage
Conference room pendents	Peerless LD7-010450	1-F32-T8-SP30 (32-watt, 3000°K) triphosphor by GE in cross section	Lutron Hi-Lume FDB-4827-120-2 (dimmable electronic with range from 1% to 100% with the specified lamp)	67 watts for 2 lamps/1 ballast
Conference room downlights	Kurt Versen P625-HP-W (Wheat cone for best incandescent-like appearance of compact fluorescent downlight)	1-F18DBXT4-SPX27 (18-watt, 2700°K) double compact triphosphor by GE operated in vertical orientation for best incandescent-like appearance of downlight)	Kurt Versen-supplied high-power-factor electromagnetic	21 watts for 1 lamp/1 ballast
Conference room accent lights	Kurt Versen C7309-G	1-50PAR30-HIR-NFL 25° halogen infrared lamp by GE	N/A	50 watts

lighting, the long-life, high-efficiency halogen infrared PAR30 lamps were recommended. Finally, photo sensor, occupancy sensor, and time controls were recommended to minimize energy use when users were not in the workspaces.

Table 4-4 summarizes the specific lighting equipment recommendations for the Steelcase Corporate Headquarters Project. The recommendations resulted in a visual environment that now meets illuminance, luminance, and luminance ratio criteria for electronic workplaces. Further, high-color-rendering triphosphor lamps provide vivid chromatic contrast. Finally, an aggregate power budget of 16 w/m^2 was achieved, which is a 22 percent reduction from preretrofit conditions. If the effects of controls are included, the "effective" connected load of the retrofit is about 12.8 w/m^2, or a 37 percent reduction from preretrofit conditions.

There are some rather subtle differences in criteria priorities between this project and The MacArthur Foundation Headquarters Project (see *Case Study 1*). Steelcase is concerned about the *earth-friendly* issues, so much so that issues regarding initial costs on efficient lamps, ballasts, and luminaires were not questioned. Nevertheless, Steelcase does have a concern regarding maintenance staffing, particularly relevant given its physical plant size versus The MacArthur Foundation. As such, Steelcase was concerned about lamp replacement cycles—the longer the better to minimize maintenance staffing. This led to the use of *rapid start* electronic ballasts to softly start fluorescent lamps, resulting in about twice as much lamp life as *instant-start* electronic-ballasted lamps. The penalty, however, for rapid-start electronic ballasts over instant-start electronic ballasts is about a 3 percent energy increase. That is, although lamp life is improved

> Lamp life nearly doubles with rapid start ballasts and lamps compared to instant start ballasts and lamps. Although instant ballasts and lamps offer a 3 percent energy reduction, the shorter lamp life may negate these savings.

with rapid start electronic ballasts, energy consumption is about 3 percent higher than with instant start electronic ballasts. Such level of detail and balancing of criteria are crucial to developing acceptable solutions for each client/project.

CASE STUDY 4

Gary Steffy Lighting Design Inc. Studio

Project Credits

Project: Gary Steffy Lighting Design Inc. Studio
Client/Owner: Gary Steffy Lighting Design Inc.
Interior Architect: Gary Steffy Lighting Design Inc.
Architect of Record: DeConti/Jernigan and Associates, Inc.
Lighting Design: Gary Steffy Lighting Design Inc.
Photos: Fred Golden

Introduction

Gary Steffy Lighting Design Inc. (GSLD) has operated a studio since 1982. In 1993 the decision was made to move to a larger space that simultaneously offered improved rental rates and operating expenses. This provided the challenge of developing a visually exciting and unique space, while remaining true to lighting criteria appropriate for CADD tasks.

Recommended Lighting Criteria

The lighting criteria outlined for this project and used to develop design solutions were different in some aspects than criteria used for The MacArthur Foundation Headquarters, The Prudential Network Command Center, and the Steelcase Corporate Headquarters. Although intensive VDT use was expected to occur, the nature of this work was expected to vary from CADD VDT tasks to conversational VDT tasks and data-entry VDT tasks. Additionally, meeting with people was anticipated as likely.

Illuminances

Given the firm's software and hardware, in many cases negative-contrast VDT screens would be available. Paperwork, from reading handwritten notes to scaled blueprints, would likely range from high contrast to low contrast. Therefore, ambient or general lighting levels throughout the studio area could range from 225 to 565 lux average maintained at work surface height (see Table 3-1). At the specific zone where

> The design challenge was to develop a visually exciting and unique space yet remain true to lighting criteria appropriate for CADD tasks.

paperwork occurs, lighting levels should not exceed 750 lux (total, including ambient lighting) and, preferably, should approach 500 lux to minimize contrast and resulting transient adaptation when viewing between paper documents and VDTs. To avoid transient adaptation effects that might occur when viewing from one surface or area to another surface or area, the minimum illuminance should be about 200 lux at work surface height. To limit the possibility of VDT screen washout, vertical illuminances at 1.2-meters AFF in the four primary viewing directions (e.g., north, south, east, and west) should be less than 250 lux.

Luminances

Maximum luminance of any surface in the environment should be less than 850 cd/m^2 and preferably 510 cd/m^2.

Luminance Ratios

Luminance ratios of 5:1 (maximum to minimum) or less from area to area or surface to surface are necessary if significant reduction and/or elimination of unwanted glare and reflections in VDT screens is desired. This suggests that light level uniformity from one area or surface to another is important. Similarly, surface reflectances must not vary wildly from one another.

Luminaires

Luminaire luminances should not exceed 850 cd/m^2 at 55°, 340 cd/m^2 at 65°, and 170 cd/m^2 at 75° (see Figure 3-1).

Desk-mounted and/or binder bin-mounted task lights should provide the user with some sort of user control, either physical adjustment in height, orientation or location over the task surface, and/or light level control via dimmer or multilevel switch mounted on the luminaire.

Electronic ballasts should be used to avoid annoying hum and flicker, as well as to maximize efficient operation of the lighting system.

Surface Reflectances

As indicated previously, wild variations in surface reflectances should be avoided. All surfaces should have a matte finish to limit glary reflections. Ceiling reflectance should be near 80 percent. Wall reflectances should range between 30 percent and 50 percent. This includes the partial-height workstation partitions. Floor reflectances should be near 20 percent. Work surfaces should have reflectances ranging between 20 percent and 40 percent. Table 3-2 in Chapter 3 should be referenced for additional surface reflectance information.

Power Budget

Although the ASHRAE/IES 90.1—1989 Standard indicates that 20 w/m² is acceptable, given previous experiences (see *The MacArthur Foundation Headquarters*

Case Study), a power budget of no more than 16 w/m² is established as a design target.

Psychological Aspects

To achieve the luminances and luminance ratios discussed previously, many times the environmental setting is visually bland. To minimize this bland sense, some accent lighting is desirable. Additionally, wall lighting or vertical accents help provide a relatively more spacious and relaxed setting.

Improved chromatic contrast seems to result in an impression of increased brightness. The newer, more efficient triphosphor fluorescent lamps offer improved color rendering and thus provide an impression of increased brightness. Warm-toned lamps seem to provide a softer, less tense setting, perhaps enhancing more interpersonal relationships or interactions. Lamps with high color rendering and warm-toned color should be considered.

Demonstration Capabilities

As a design studio, inherent expectations are held by those clients who visit the studio. Demonstrating lighting effects, equipment hardware options, controls, and the like are quite illustrative in convincing visitors of appropriate techniques. Additionally, the fact that "what one preaches is what one practices" strengthens credibility. This criterion was deemed important in developing lighting solutions and space plans.

Recommended Lighting Approach

A rectilinear shell space of about 6.7 m by 16.8 m by 5.5 m in height was secured, with a plan to paint all surfaces a high reflectance white and drop a partial ceiling at 3.6 meters above the finished floor. White surfaces were used to maximize lighting distribution efficiency. Since the all-white backdrop would be visually bland, uplighting of the upper deck (essentially lighting the plenum space from 3.6 m to 5.5 m) in a deep sky-blue was introduced. Recognize that this provides chromatic contrast while actually assisting in luminance contrast (the upper plenum won't be a black hole against other lighted surfaces; such a black hole would cause luminance contrast problems and lead to veiling reflections on the VDT screens as well as lead to transient adaptation problems as users scanned the room). Figure 4-13 illustrates the luminance provided by the uplighting of the upper deck.

Daylighting is minimal, with a northern exposure glass door wall. Workstations were configured to minimize the impact of daylight luminances. Additionally, surface luminances were addressed throughout the workspace to help balance the daylight luminances.

The linear space lends itself to a definitive segregation of tasks, with the more laborious VDT and drafting tasks along one of the long walls and the reference

FIGURE 4-13

The suspended ceiling is left partially incomplete so that a view of the blue-uplighted plenum is available to users of the space. The uplighting in the plenum (from the cove on top of the library wall on the right) also helps balance luminances, important for VDT use and to minimize transient adaptation effects. Courtesy of Fred Golden.

library along the opposite wall. As such, lighting could be oriented specifically for each of these task areas, rather than spreading lighting uniformly throughout the space. This can result in significant energy reductions when compared to the more traditional approach of providing uniform lighting throughout a space.

For the VDT and drafting task area, the dropped ceiling height was set at 3.6 meters to both allow for a more uniform diffusion of indirect lighting as well as enable the use of direct/indirect equipment without introducing too much light and/or glare from the direct component. Direct/indirect equipment can be some of the most efficient lighting hardware available, trapping very little light. Though a variety of completely indirect approaches were considered, the dropped ceiling height of 3.6 meters and the desire to provide the option of ambient lighting levels approaching 565 lux for drafting tasks of low contrast led to the conclusion that a direct/indirect approach met illuminance, luminance, and energy criteria. Figure 4-14 illustrates the studio area.

To provide fill lighting for the drafting task area, to balance luminances of the surrounding surfaces, and to adequately light the reference library, low-wattage, small-scale fluorescent wall wash lighting was used. This provides a uniform wash of light from floor to the 3.6 meter height of the dropped ceiling. Figure 4-15 shows the hardware and effects of the wall wash lighting.

In the conference area, a series of small-scale pendents defines the conference setting. Accent lighting is used to provide additional general lighting on the conference table for those times when reviewing of written and drawn material is necessary. Additionally, accent lighting is introduced onto an enlarged photograph illustrating the firm's work and providing a dramatic backdrop to the conference area.

FIGURE 4-14

The VDT/drafting studio area is lighted with direct/indirect pendent luminaires on 1.8 meter spacings that are oriented at the work area to minimize energy use typical of more uniform lighting layouts. Luminances, however, were not sacrificed. The use of appropriate indirect optics and suspension heights, the use of background lighting along the reference library, and the use of light colored surfaces were necessary to meet luminance and luminance ratio criteria. Courtesy of Fred Golden.

FIGURE 4-15

Small-scale, low-wattage compact fluorescent wall wash lights are used to uniformly light the reference library wall. This helps with task lighting of the reference books, with luminance balancing of the entire studio, and with impressions of spaciousness. Courtesy of Fred Golden.

This accent also helps in introducing a sense of relaxation and spaciousness as well as helping to balance luminances from the conference table tasks to the background. Finally, to add sparkle and visual interest to the remainder of the conference area, small-scale, low-wattage steplights are used on close spacings. Figure 4-16 shows the conference area.

Findings: Studio Work Area

The following combinations of equipment meets or exceeds the previously described criteria:

▼ ambient lighting: a Peerless Rounded Mostly Up direct/indirect luminaire suspended 500 mm below the ceiling (clear) and of 2.5 m in length by 85 mm in depth and 250 mm in width with two lamps in cross section (a GE SP30 triphosphor T8 fluorescent F32T8/SP30) along with Lutron Hi-Lume electronic dimming ballasts (one ballast serving two lamps)

▼ architectural fill lighting: a Williams asymmetric strip luminaire mounted in a cove at 3.6 m above finished floor with two lamps in cross section (a GE blue T12) along with Lutron Hi-Lume electronic dimming ballasts (one ballast serving two lamps)

▼ library wall wash lighting: a Columbia Parawash luminaire of 200 mm in width by 300 mm in length and 100 mm in depth with one rapid start 18-watt compact fluorescent lamp (a GE F18BX/SPX30/RS) along with Robertson rapid-start electronic ballasts (one ballast per one lamp)

FIGURE 4-16

The conference area has visual impact with saturated red pendents, the sparkle of the step lights in the background, and the accent of the mounted photograph. Courtesy of Fred Golden.

TABLE 4-5   *Recommended Lighting Equipment Summary (For Gary Steffy Lighting Design Inc. Studio)*

Application	Luminaire	Lamp	Ballast	Wattage
Ambient lighting system	Peerless LD3-320501	2-F32-T8-SP30 (32-watt, 3000°K) triphosphor by GE in cross section	Lutron Hi-Lume FDB-4827-120-2 (dimmable electronic with range from 1% to 100% with the specified lamp)	67 watts for 2, 32-watt lamps/1 ballast
Architectural fill lighting system	Williams 7422-R1348	2-F40T12/Blue rapid start lamps by GE in cross section (staggered arrangement to avoid socket-shadow)	Lutron Hi-Lume FDB-4843-120-2 (dimmable electronic with range from 1% to 100% with the specified lamp)	83 watts for 2, 40-watt lamps/1 ballast
Reference library wall wash lighting system	Columbia Parawash LWW-1-S-49-1-1-120	1-F18BX-SPX30RS (18-watt, 3000°K) compact triphosphor by GE	Robertson RER1LT8-120R	20 watts for 1, 18-watt lamp/1 ballast
Accent lighting system	Litelab Max-Track with L-115-S10-P Luminaires (3 in Studio and 2 in conference area)	1-50PAR30/HIR/NFL 25° for each of 3 in Studio and 1-50PAR30/HIR/NSP7° for each of 2 in Conference Area by GE	N/A	50 watts
Task lighting system (at selected work surfaces)	Luxo LC	1-22-watt FC8T9-SW (22-watt, 3025°K) circline fluorescent by GE and 1-50TB/H (50-watt tungsten-halogen) by GE	Luxo-supplied low-power-factor electromagnetic	80 watts for 1-20-watt lamp/1 ballast and 1-50-watt lamp
Conference area pendents	Leucos Golf Satin Red	1-Q50MR16/FL40° (50-watt, 12-volt) by GE	Leucos-supplied low-voltage electromagnetic transformer	60 watts for 1 lamp/1 transformer
Conference area steplights	Bega 1123	1-Q20T3/CL (20-watt, 12-volt, clear) by GE	Bega-supplied low-voltage electromagnetic transformer	25 watts for 1 lamp/1 transformer

▼ accent lighting: a Litelab MaxTrack (busway track system) mounted at 5 m above finished floor with three Litelab L-115 luminaires (115 mm in diameter and 165 mm in length); and each luminaire using one 50-watt PAR30 halogen infrared narrow flood lamp (a GE 50PAR30/HIR/NFL25°).

▼ task lighting: in some instances, depending on user preference, a freestanding task light is required, in which case standard Luxo drafting lights are used, which have two lamps—a 50-watt tungsten-halogen A-lamp (GE 50TB/H) and a 22-watt circline soft-white "Home" fluorescent lamp (GE FC8T9-SW). The soft-white color is an improvement over the typical cool-white lamps available in this type of lamp.

When this equipment is combined, the overall luminances and illuminances are quite appropriate for the VDT and paper tasks involved. Additionally, the dimming enables users to set work scenes for extensive (in either or both duration and detail) drafting tasks. Further, the chromatic contrast introduced by the blue uplighting within the exposed plenum provides visual interest and character typically not found in commercial work settings.

Findings: Conference Area

The conferencing area is located at one end of the space, which looks into a mall. To introduce lighting that was distinctively appropriate with conferencing functions and to provide visually interesting focus for mall-goers, a series of very small scale, saturated red pendents was used over the table. High above the table, a track has two 50-watt halogen infrared narrow spot lights—one that provides additional table lighting and one that provides a strong focus on a photograph. Since the photograph and pendents are visible from the studio area, they act as visually distant foci that help lead visitors through the studio as well as provide a "resting" place for users' eyes, which might otherwise remain accommodated to close VDT tasks.

Summary

All of the lighting components previously described, when used in conjunction with a Lutron Grafik Eye 3000-series preset scene control, provide a general work setting and conferencing lighting arrangement that has an effective connected load of 15.4 w/m^2 (many lights are dimmed to between 30 and 90 percent full output). Without the effect of controls, the actual connected load is 21 w/m^2, which in and of itself is quite remarkable for such a small space housing such demanding visual tasks. The controls, however, reduce the actual connected load by 27 percent yet offer flexibility to adjust lighting for varying tasks and/or for lumen depreciation caused by lamp aging and dirt accumulation.

Table 4-5 summarizes the specific lighting equipment recommendations for the Studio of Gary Steffy Lighting Design Inc.

Reference and Product Resources

There are many manufacturers of fine lighting equipment. Many of the design and engineering industry publications offer annual product directories that are comprehensive—perhaps most notably, the Lighting Equipment & Accessories Directory in the February issue of *Lighting Design & Application* Magazine (120 Wall Street, 17th Floor, NYC, NY 10005-4001). This text referred to a variety of specific equipment in Chapter 4, *Using the Model Guideline: Case Studies*. For additional information on referenced equipment, contact the manufacturer. Following is a list of manufacturers and organizations referenced in this text. A "v" indicates "voice-contact" phone number; "f" indicates "facsimile-contact" phone number. These addresses and phone numbers were current effective the first quarter of 1995.

Advance Transformer Company
10275 West Higgins Road
Rosemont, Illinois 60018
v/708-390-5000
f/708-390-5109

Alkco
11500 West Melrose Avenue
Franklin Park, Illinois 60131
v/708-451-0700
f/708-451-7512

Artemide Inc.
1980 New Highway
Farmingdale, New York 11735
v/516-694-9292
f/516-694-9275

Bega
P.O. Box 1285
Carpinteria, California 93014-1285
v/805-684-0533
f/805-684-6682

Chartered Institution of Building Services Engineers (CIBSE)
(*Lighting Guide: Areas for Visual Display Terminals*, LG3: 1989)
Delta House
222 Balham High Road
London SW12 9BS
United Kingdom

Columbia Lighting
P.O. Box 2787
Spokane, Washington 99220
v/509-924-7000
f/509-924-1134

Deutsches Institut für Normung (DIN)
(*Artificial Lighting of Interiors: Lighting of Rooms with VDU Workstations or VDU Assisted Workplaces*, 1988)
American National Standards Institute
11 West 42nd Street
New York, New York 10036
v/212-642-4900

General Electric Lighting
Nela Park
1975 Noble Road
Cleveland, Ohio 44112-6300
v/800-626-2004

H.E. Williams Inc.
P.O. Box 837
Carthage, Missouri 64836-0837
v/417-358-4065
f/417-358-6015

Illuminating Engineering Society of North America (IESNA)
(*IES Recommended Practice for Lighting Offices Containing Computer Visual Display Terminals*, 1990)
120 Wall Street
17th Floor
New York, New York 10005
v/212-705-7926

International Commission on Illumination (CIE)
(*Vision and the Visual Display Unit Work Station*, 1984)
United States National Committee
ARC Sales
7 Pond Street
Salem, Massachusetts 01970-4893
v/508-745-2249

Kurt Versen
10 Charles Street
Westwood, New Jersey 07675
v/201-664-8200
f/201-664-4801

Leucos USA, Inc.
70 Campus Plaza II
Edison, New Jersey 08837
v/908-225-0010
f/908-225-0250

Litelab Corporation
251 Elm Street
Buffalo, New York 14201
v/800-238-4120
f/716-856-0156

Lithonia Lighting
1335 Industrial Boulevard
Conyers, Georgia 30207
v/404-922-9000
f/404-922-1841

Louis Poulsen/See Poulsen Lighting

Lutron Electronics Co., Inc.
7200 Suter Road
Coopersburg, Pennsylvania 18036
v/215-282-3800
f/215-282-6431

Luxo Corporation
36 Midland Avenue
Port Chester, New York 10573
v/914-937-4433
f/914-937-7016

MagneTek
200 Robin Road
Paramus, New Jersey 07652
v/201-967-7600
f/201-967-0904

Motorola Lighting Inc.
887 Deerfield Parkway
Buffalo Grove, Illinois 60089
v/800-654-0089
f/708-215-6444

Neo-Ray Lighting Systems Inc.
537 Johnson Avenue
Brooklyn, New York 11237
v/718-456-7400
f/718-456-5492

OSRAM Sylvania Inc.
100 Endicott Street
Danvers, Massachusetts 01923
v/508-750-2290
f/508-750-2089

Peerless Lighting Corporation
2246 Fifth Street
Berkeley, California 94710
v/510-845-2760
f/510-845-2776

Poulsen Lighting, Inc.
5407 Northwest 163rd Street
Miami, Florida 33014-6130
v/305-625-1009
f/305-625-1213

Robertson Transformer Company
13611 Thornton Road
Blue Island, Illinois 60406
v/708-388-2315
f/708-388-2420

Ron Rezek/See Artemide Inc.

Steelcase Inc.
P.O. Box 1967
Grand Rapids, Michigan 49501
v/616-247-2710

Sylvania/See OSRAM Sylvania

Watt Stopper, Inc.
296 Brokaw Road
Santa Clara, California 95050
v/800-879-8585
f/408-988-5373

Williams/See H.E. Williams Inc.

General References

Alexander, Michael. "Redesigning the Office of the Future." *Computerworld*, April 2, 1990, p. 18.

American Society of Heating, Refrigerating and Air-Conditioning Engineers, Inc. and Illuminating Engineering Society of North America. *ASHRAE/IES 90.1–1989: Energy Efficient Design of New Buildings Except New Low-Rise Residential Buildings.* Atlanta: ASHRAE/IES, 1989.

Attwood, Dennis. "Comparison of Discomfort Experienced at CADD, Word Processing and Traditional Drafting Workstations." *International Journal of Industrial Ergonomics*, 4 (July, 1989), pp. 39–50.

California Energy Commission, *P400-88-005 Energy Efficiency Manual: Designing for Compliance, Second Generation Nonresidential Standards*, 3rd ed. Sacramento: CEC, 1988.

Flynn, John E. "A Study of Subjective Responses to Low Energy and Nonuniform Lighting Systems." *Lighting Design & Application*, February, 1977, pp. 6–14.

Flynn, John E., Kremers, Jack A., Segil, Arthur W., and Steffy, Gary R. *Architectural Interior Systems*, 3rd ed. New York: Van Nostrand Reinhold, 1992.

Goldman, Jerome, and Aldich, Franklin. "Optical Discomfort from VDTs and Fluorescent Lighting (letter)." *Journal of the American Medical Association*, 264 (September 5, 1990), p. 1174.

Hedge, Alan. "Lighting the Computerized Office: A Comparative Field Study of a Lensed Indirect Uplighting System and a Parabolic Downlighting System." Unpublished report, Cornell University, 1989.

Loach, Kenneth. "How to Achieve Effective Lighting in Your Offices." *The Office*, December 1989, pp. 65ff.

Tijerina, Louis. *Ergonomic Guidelines for VDT Workstation Design.* Dublin, Ohio: OCLC OnLine Computer Library Center, Inc. 1985.

Travis, David, Bowles, Susan, Seton, John, and Peppe, Roger. "Reading from Color Displays: A Psychophysical Model." *Human Factors*, 32 (April, 1990), pp. 147–156

Watson, Robert K. "Case Study of Energy Efficient Building Retrofit: 40 West 20th Street, Headquarters of the Natural Resources Defense Council." New York: Natural Resources Defense Council, Inc., March 1990.

Welch, Mark J., and Bater, Barry. "VDTs and Health: Just the Facts." *Legal Times*, October 16, 1989, pp. 53–55.

Yearout, Robert, and Konz, Stephan, "Visual Display Unit Workstation Lighting." *International Journal of Industrial Ergonomics*, 3 (April, 1989), pp. 265–273.

Endnotes

1. Nathan Shedroff, J. Sterling Hutto, and Ken Fromm, *Understanding Computers*, Alameda, CA: SYBEX, 1992.

2. Michael Hammer and James Champy, *Reengineering the Corporation*, New York: HarperCollins Publishers, Inc. 1993.

3. Frank Bryant, "VDT Debate Centers on Direct vs. Indirect Lighting," *Energy User News*, May, 1991, pp. 26–27.

4. Mark S. Rea, ed., *Lighting Handbook Reference & Application.* 8th ed., New York: Illuminating Engineering Society of North America, 1993.

5. P. R. Boyce, *Human Factors in Lighting*, New York: Macmillan Publishing Company, 1981.

6. Gary R. Steffy, *Architectural Lighting Design*, New York: Van Nostrand Reinhold, 1990.

7. International Commission on Illumination, *Vision and the Visual Display Unit Work Station*, Paris: International Commission on Illumination, 1984, p. 12.

8. International Organization of Standardization Technical Committee 159, Subcommittee 4, *Ergonomic Requirements for Office Work with Visual Display Terminals (VDTs), Part 6, Environmental Requirements*, Committee Draft ISO 9241–6, December 13, 1990, p. 23.

9. International Commission on Illumination, *Vision and the Visual Display Unit Work Station*, Paris: International Commission on Illumination, 1984, p. 14.

10. O. D. Godnig, C. Edward, and John S. Hacunda, *Computers & Visual Stress*, Grand Rapids, MI: Abacus, 1991.

11. Wilbert O. Galitz, *User-Interface Screen Design*, Wellesley, MA: QED Information Sciences, Inc., 1993.

12. Public Works Canada, *Office Lighting for Video Display Terminals*, Draft, Public Works Canada, Ottawa, 1987, pp. 7, 12.

13. George C. Brainard, "The effects of light on physiology, mood, and behavior in humans," in *Psychological Aspects of Architectural Lighting Symposium Proceedings*, 90–98. The Pennsylvania State University, Department of Architectural Engineering. University Park, PA, 1990.

14. M. Terman and R. McCluney, "Counteracting daylight deprivation," in *Commission Internationale de l'Eclairage Proceedings of 21st Session, Venice, Volume I, Publication 71*, (CIE: Vienna, 1987), 346–349.

15. American National Standards Institute, *American National Standard for Human Factors Engineering of Visual Display Terminal Workstations* (Santa Monica: The Human Factors Society, 1988).

16. Deutsches Institut für Normung, *DIN 5035 Part 7: Artificial Lighting of Interiors: Lighting of Rooms with VDU Workstations or VDU Assisted Workplaces* (Berlin: Deutsches Institut für Normung, 1988).

17. New Jersey Department of Health Public Employees Occupational Safety and Health Program *Guidelines for the Use and Functioning of Video Display Terminals, Part I* (Trenton: New Jersey Department of Health, 1989).

18. Illuminating Engineering Society Office Lighting Committee Subcommittee on Visual Display Terminals, *IES Recommended Practice for Lighting Offices Containing Visual Display Terminals* (RP-24) (New York: Illuminating Engineering Society of North America, 1990).

19. The Chartered Institution of Building Services Engineers, *Lighting Guide: Areas for Visual Display Terminals* (London: The Chartered Institution of Building Services Engineers, 1989).

20. International Commission on Illumination Technical Committee 3.1 (Visual Performance), *Vision and the Visual Display Unit Work Station* (CIE-60) (International Commission on Illumination, 1984).

21. Chartered Institution of Building Services Engineers, *Technical Memoranda TM6: Lighting for Visual Display Units* (London: Chartered Institution of Building Services Engineers, 1981).

22. John E. Flynn, "A Study of Subjective Responses to Low Energy and Nonuniform Lighting Systems." *Lighting Design & Application*, February, 1977, pp. 6–14.

23. Richard J. Farrell and John M. Booth, *Design Handbook for Imagery Interpretation Equipment*, (Seattle: Boeing Aerospace Company, 1984).

24. International Organization of Standardization Technical Committee 159, Subcommittee 4. *Ergonomic Requirements for Office Work with Visual Display Terminals (VDTs), Part 6, Environmental Requirements*, Committee Draft ISO 9241–6 (Geneva, December 13, 1990).

25. Public Works Canada, *Office Lighting for Video Display Terminals*, Draft (Ottawa: Public Works Canada, 1987).

26. City of San Francisco, *Health Code of the San Francisco Municipal Code, Part II, Chapter 5, Article 23* (San Francisco: City of San Francisco, 1990). [Note: Ordinance declared unconstitutional in 1991.]

Survey Forms

The reader may use the following forms on a project-by-project basis for data collection and design. Mass reproduction and/or reproduction for sale or distribution is prohibited. Copyright 1995, Gary R. Steffy.

Visual Task Survey

Visual Tasks	Anticipated Frequency			Anticipated Importance			Comments
	Lots	Fair Amount	Not Much	Great	Moderate	Little	
☐ Reading handwriting in ink	☐	☐	☐	☐	☐	☐	
☐ Reading handwriting in pencil	☐	☐	☐	☐	☐	☐	
☐ Reading printed matter							
☐ Small	☐	☐	☐	☐	☐	☐	
☐ Large	☐	☐	☐	☐	☐	☐	
☐ High contrast (crisp, black)	☐	☐	☐	☐	☐	☐	
☐ Low contrast (fuzzy, gray)	☐	☐	☐	☐	☐	☐	
☐ Writing in ink	☐	☐	☐	☐	☐	☐	
☐ Writing in pencil	☐	☐	☐	☐	☐	☐	
☐ Facial recognition	☐	☐	☐	☐	☐	☐	
☐ Accounting	☐	☐	☐	☐	☐	☐	
☐ Ledgers	☐	☐	☐	☐	☐	☐	
☐ Currency	☐	☐	☐	☐	☐	☐	
☐ Video display terminal/CRT[a]							
☐ Monochrome	☐	☐	☐	☐	☐	☐	
☐ Color	☐	☐	☐	☐	☐	☐	
☐ Video display terminal/LCD[b]	☐	☐	☐	☐	☐	☐	
☐ Other/describe	☐	☐	☐	☐	☐	☐	

[a]Cathode ray tube technology provides an internally lighted screen, much like a television.
[b]Liquid crystal diode technology requires external lighting in order to provide a visible screen.

Existing Lighting Survey, Part 1

Building ID   Date   Time

Window
☐ YES   Location   N E S W
☐ NO

Window Treatment
☐ YES   Position   Open Shut Part
☐ NO

Sky Condition   Clear   Hazy Clear   Partly Cloudy   Mostly Cloudy   Direct Solar Disk

Ambient lighting conditions
Luminaires   No Shielding   Painted Louvers   Baffles   Lens (what kind)   Wraparound
Recessed   Surface Mounted   Pendent   Condition (maintenance)
Lamps   Type   Wattage   Quantity/luminaire
Controls   Local by user   General by supervisor

Task lighting conditions
Luminaires
Lamps
Controls

Visual Tasks
☐ Reading handwriting   pen   pencil
☐ Reading printed matter   small   large   high contrast   low contrast
☐ Writing   pen   pencil
☐ Facial recognition
☐ Accounting   ledgers   currency
☐ Video display   B+W   color   Screen characteristics
☐ Other/describe

Existing Lighting Survey, Part 2

Workstation colors
- ☐ Work surface
- ☐ Partitions if applicable
- ☐ Floor
- ☐ Files
- ☐ Walls
- ☐ Ceiling

Comments

Photos

Illuminances
- ☐ Primary work surface
  [insert or sketch a plan layout of work area and mark illuminances]
- ☐ Secondary work surface   Vertical work surface (VDT)

Luminances
- ☐ *Work Surface*
- ☐ *Paper task(s)*
- ☐ *VDT*
- ☐ *Window*
- ☐ *Ceiling*
- ☐ *Wall(s)*
- ☐ *Luminaire(s)*
  - ☐ *N*
  - ☐ *E*
  - ☐ *S*
  - ☐ *W*
- ☐ *Partitions*
- ☐ *Partition under overheads*
- ☐ *Binder bins/overheads*
- ☐ *Other*

Index

References to pages with related figures are in italics type. References to pages with related tables are in bold type.